Thermal Imaging for Wildlife Applications

Thermal Imaging for Wildlife Applications

A Practical Guide to a Technical Subject

Kayleigh Fawcett Williams

DATA IN THE WILD SERIES

Pelagic Publishing

First published in 2024 by
Pelagic Publishing
20–22 Wenlock Road
London N1 7GU, UK

www.pelagicpublishing.com

British Library Cataloguing in Publication Data
A catalogue record for this book is
available from the British Library

ISBN 978-1-78427-416-0 Hbk
ISBN 978-1-78427-387-3 Pbk
ISBN 978-1-78427-388-0 ePub
ISBN 978-1-78427-389-7 PDF

https://doi.org/10.53061/IPTU9371

Cover image: Fallow Deer by Kayleigh Fawcett Williams. Taken using FLIR T1030sc.

Contents

Figures

Preface

High-quality, reliable data is needed now more than ever before to inform decisions that will ultimately shape the future biodiversity of our planet. Thousands of humans, including ecologists, biologists, researchers, wildlife managers and more, are working to better understand and protect wildlife globally. In light of the fact that you are reading this book, I suspect that you may be one of them. As a wildlife professional (or dedicated amateur recorder), you and/or your organisation most likely require some kind of data to inform the work that you do. The type and quality of that data you collect will impact decisions, actions and resulting outcomes for wildlife (and humans too). What you put in has a big influence on what you get out. Therefore, positive outcomes rely heavily on the acquisition of appropriate, accurate and reliable data. The successful application of thermal imaging in our work is a powerful way to maximise the quality of our inputs, so as to facilitate vast improvements in our outputs.

When applied appropriately this technology can give us the power to dramatically improve the data we collect and, as a result, the effectiveness of the actions that data is often sought to inform. As a technical discipline in its own right, thermal imaging is not the simplest of technologies to apply, and it can bring along a suite of benefits and challenges to our wildlife work which you can begin to learn more about in Chapter 1.

My aim in writing this book is to bring to you a practical guide to a technical, and often inaccessible subject. To do this, I will share input from my own technical knowledge as a thermographer alongside my experience as a wildlife professional. In Chapter 2, Foundations, I will outline the key concepts we need to understand to bring this technology and wildlife applications together effectively.

Whatever the wildlife application, good methods are central to the collection of reliable and accurate data, and we'll explore these in Chapter 3. Using thermography in my wildlife work has shown me just how much we can miss when using traditional methods alone, and how powerful the use of this technology can be to avoid the pitfalls and limitations they can hold. Thermal imaging is not a silver bullet, however, and it does require a level of expertise to be used effectively. So if you do choose to go down this path, be sure to not only arm yourself with the appropriate equipment but also with the skills and knowledge you need to use that equipment effectively. Skipping this step almost always leads to errors that may go unnoticed, but can have very real consequences further down the line.

Every wildlife professional requires a toolkit for their work: yours will be different from mine. As for any task, we need access to the right tools for the job that we wish to do. If we want to carry out our jobs as effectively and efficiently as possible, we need to make sure that our toolkit has those tools in it, that we know which tools to select for each task and, once selected, how to use them properly. As a global community of wildlife professionals, we need access to a library of different tools for the variety of challenges we face. In this book, I want to show you what a valuable tool thermal imaging can be. Wildlife-related tasks come with inherent challenges, and there is no one-size-fits-all technology out there to solve all of them. The menu of options for tech-based solutions to the challenges we face grows year on year, and this double-edged sword provides us with both an attractive array of options and a paradox of choice. Implementing a tech-based solution successfully requires the selection of the most appropriate technology (or more often a combination of technologies) for our needs. Thermal imaging is no exception: in fact, I would argue that it is especially important to choose this type of equipment very carefully. In Chapter 4, we will explore this process to maximise our potential for success.

The research for this book has involved the review of over 300 documents. In Chapters 5 and 6, I explore a range of applications that I have read about during my extensive reading of scientific literature, as well as others that are not well represented in the published research. Here we break the applications down into broad application categories and species groups. Although these different categories of wildlife-related applications may use the same core technology, they can require very different device specifications, methods and expertise as we explore in later chapters.

Technology is evolving. By the time you read this book, technology for you will be more advanced than that available for me to write about in 2022. From the latter perspective, in Chapter 7 I consider what likely changes we might expect to see and crucially what is needed for the more widespread and efficient use of thermal imaging in wildlife-related fields in the future. I hope you'll keep looking forward to more and more possibilities from there.

How to use this book

I have structured this book in such a way that it may be read from cover to cover or be used as a reference. To help you find the information you need, I have created individual chapters broken down into sub-headed sections which can be read separately. Where relevant, I will direct you to pertinent information within other sections or chapters of this

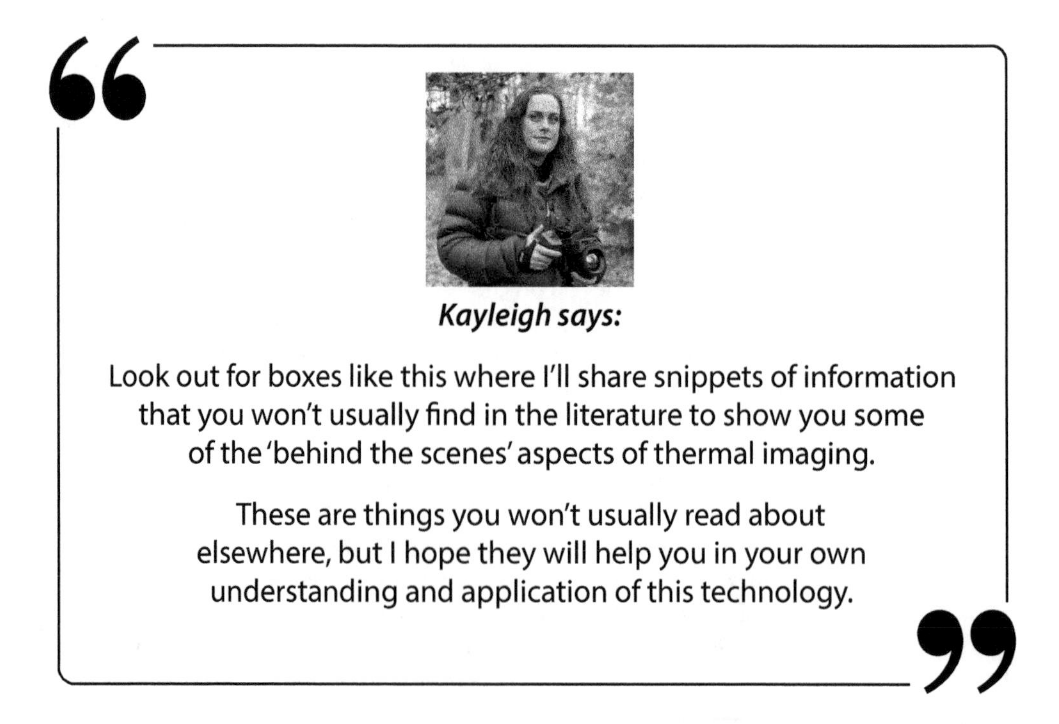

book, and also to other key texts available. You'll find case studies provided throughout the book to help illustrate real-life methods, applications and concepts. Finally, I have also included hints, tips and examples from my own personal experience in boxes like this throughout the book:

In descriptions of thermal-imaging studies I include, where possible, details of the thermal-imaging device and how it has been used. In general, I first include the *make* (e.g. FLIR) followed by the *model* (e.g. T1030sc) and then the *manufacturer* details in parentheses (e.g. FLIR Systems, Inc., Wilsonville, Oregon, USA). You can also find a breakdown of make and model of all equipment used in the studies I have reviewed in the tables contained in the Appendix.

My story

My thermal-imaging and wildlife journey began when I started my PhD back in 2010. I arrived at the University of Southern Denmark around the same time as a shiny new high-speed thermal-imaging camera (a FLIR SC5000). My doctoral thesis focused on bat bioacoustics and involved a lot of acoustic technology. Yet, despite the complex multi-microphone recording systems I already had to deal with for my research, I was incredibly curious about the possibility of using this visual gadget that had arrived in our lab. Before long, I was lucky enough to be able to try it out. After many failed attempts in rapid succession, I quickly learned that this was not going to be a case of turning the thing on and hitting the record button. No: if I was going to be able to gather any usable data I would have to learn how to use this properly. My supervisors kindly allowed me to return to the UK to visit FLIR HQ where I would spend a week training to be a 'Thermographer'. The Level 1 Thermography Certification course I took opened my eyes to the fascinating world of thermal imaging and got me started using the SC5000 properly.

Having completed my PhD, I came home to the UK and began working as a bat ecologist at a large international multidisciplinary consultancy. I worked there for several years, developing thermal-imaging methods for bats, and eventually was able to take my Level 2 Thermography Certification. This course blew my mind and further fuelled my fascination with the subject. It was at this point that I realised that this technology had the potential to dramatically change the way we work with wildlife. I knew I could help make that happen, but I also knew that I couldn't do it from where I sat in my conventional ecology consulting role.

So in 2017, I took the terrifying leap to set up my own business so I could help others to survey, monitor and protect wildlife using thermal-imaging technology. Since then, I have written the *Thermal Imaging: Bat Survey Guidelines* in association with the Bat Conservation Trust (published in 2019 and updated in 2021), developed training courses and programmes for wildlife professionals, and carried out cutting-edge research and consulting work in this field.

Now, in 2022, I am using the knowledge and experience I have gained over the past 12 years of working with thermal imaging for wildlife to create this book for you. Whatever your specific wildlife interests, I hope that it helps you to improve your understanding of thermal imaging and how it might help you in your own work in the future. I hope it will inspire you to talk with others about what you learn. Together, we can use thermal imaging to help us better understand and protect a wide range of wildlife species, so that wildlife and humans can thrive on this beautiful planet we call home.

Dr Kayleigh Fawcett Williams

Acknowledgements

I am most grateful to all of my family and friends who have supported me in so many ways while I completed the almighty task of writing this book.

I would like to thank the wonderful people at the British Wildlife Centre and Cotswold Wildlife Park & Gardens for kindly allowing me to take thermal images of the incredible wildlife species you work with. Thanks to Nigel Fisher, Dr Danielle Linton, Keith Cohen and all who have made it possible for me to conduct my thermal-imaging research with wildlife in the wonderful Wytham Woods. I am extremely privileged to be able to work in such an incredible place.

Thank you so much to everyone who has helped me in my research, including: Line Faber Johannesen, Lauren Jones-Mullins, Sophie Bell, Abigail Rowley, Nick Willis and Conor Daly.

Enormous thanks to Nigel Massen, David Hawkins and all at Pelagic Publishing for working on the creation of this book with me.

1. Introduction

In the midst of a biodiversity crisis, wildlife professionals urgently need tools that can help them to gather accurate and reliable data. Wildlife ecologists, biologists, managers and researchers need this data to collectively better understand and protect wildlife all over our planet. At this critical point in time, the effective use of instruments such as thermal-imaging devices has never been so important. The growth in the global applications of these technologies in a broad spectrum of disciplines has led to the development of a growing range of thermal-imaging devices. These devices are becoming more user friendly and, as global demand drives higher production rates, economies of scale have driven down the price of the units on offer. We have now reached a point where this technology has not only become accessible to more wildlife professionals than ever before, but is increasingly becoming essential for some applications. It is important to note, however, that the successful implementation of thermal imaging for any wildlife application requires a combination of the appropriate equipment and technical expertise. Surveys should be designed to suit the specific project objectives and to overcome or minimise the likely challenges. These could be presented by the species of interest, its surrounding habitat and environmental conditions.

1.1 How it works

As a passive remote-sensing technology, thermal-imaging devices work by detecting the natural infrared waves from the surrounding environment. These electromagnetic waves (see Figure 1 overleaf) are naturally emitted and reflected by everything on our planet. Thermal-imaging devices receive these infrared waves, convert them into a digital signal, and then transform them into visual representations, known as thermal images or thermograms. These images can then be interpreted or analysed by professionals with appropriate expertise.

Thermal imaging allows for a non-invasive or non-contact approach, minimising potential for the disturbance of wildlife when gathering data. Thermal-imaging equipment can operate in all light levels, so it can be used to gain information on species with a range of different activity patterns. It has been particularly useful as a tool for use on crepuscular and nocturnal species, when light levels are low, but it can also be employed for diurnal species.

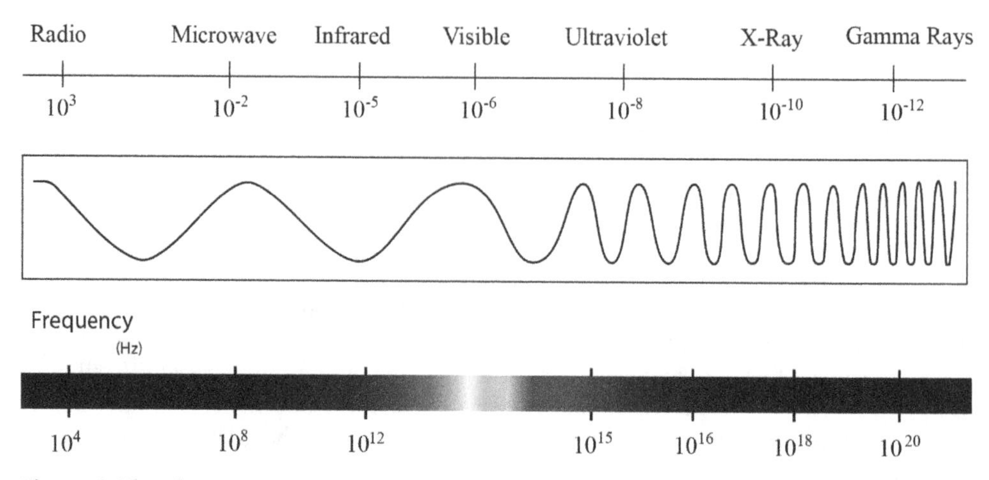

Figure 1 The electromagnetic spectrum: longer wavelengths are shown on the left-hand side and shorter wavelengths are shown on the right-hand side. Towards the centre, we find the wavelengths visible to most humans. To the left of these we find the infrared wavelengths detected by thermal-imaging devices.

Thermal Imaging vs Infrared Reflectance

Please note that we will not cover Infrared Reflectance (also often referred to as IR, Infrared, Near Infrared/NIR/Active IR) techniques in this book. However, these different technologies can sometimes be confused with one another because there are some similarities in their terminology. Therefore we will briefly clarify this before we go any further.

In contrast to the 'passive' mechanism employed by thermal imaging described above, Infrared Reflectance is an 'active' remote-sensing method, relying on the emission of IR waves which are reflected by objects in the environment. The reflected signals are received by a sensor inside the Infrared Reflectance device, which is converted into a digital signal and then into a visual representation of the scene.

1.2 History of thermal imaging for wildlife applications

At the time of writing, thermal imaging has been used for a variety of wildlife applications around the world for a select range of species, and the list is growing. According to the published literature on the subject to date (including scientific peer-reviewed journals, project reports and other articles) the list includes mammals, birds, bats, marine mammals, marsupials, insects and even fish over the past five decades or so (see Chapter 6 for more on species applications).

The published literature, however, does not tell us the full story of how this technology has been used in the field. Unfortunately, only a small fraction of the methods used on the ground for wildlife (and their subsequent results) are ever reported in such a way.

For example, thermal imaging is now commonly used for bird and bat surveys and monitoring work in the UK. These applications have been very successful, yet they receive disproportionately little attention in the scientific literature. Such work is generally only mentioned in reports that are often confidential or inaccessible to the public. Likewise, much more work has probably been happening behind the scenes in academic and applied wildlife disciplines. Though it seems likely that testing or even successful deployment of thermal imaging has been conducted in these areas, such knowledge remains largely unavailable. Historically, negative results have been hugely underreported in scientific literature, and this could be another reason why we may have an incomplete understanding of how the technology has been applied. To better comprehend how the technology is being used more widely, we will explore brief case studies on wildlife species applications of thermal imaging that are not well represented (or, in some cases, at all) in the literature.

Following the model of technology adoption laid out by Moore (1991) can help us to better understand the history, present and potential future of thermal imaging for wildlife applications. When any new technology is adopted by a market, it proceeds through several different groups of adopters in what Moore calls the 'Technology Adoption Life Cycle', in this order:

1. **Innovators** These are the first people within a market to get their hands on the technology. They want to try it out regardless of bugs, glitches and teething problems.
2. **Early adopters** This group of people are next in line to adopt the technology, but they wait until the innovators have tried it first.
3. **Early majority** The first half of people in the mainstream market to adopt the technology.
4. **Late majority** The second half of people in the mainstream market to adopt the technology.
5. **Laggards** The final group of people to adopt the technology, they wait until it has been tried and tested by the majority before they eventually adopt it.

The zone between the *early adopters* and the *early majority* is called 'the chasm': a critical gap that must be traversed for a technology to reach the mainstream. At the time of writing, we are now at the exciting point of thermal-imaging technology crossing the chasm into mainstream use for wildlife applications. The *innovators* in this field have been and gone. As the technology itself became more accessible, the *early adopters* have progressed and continue to develop methods for its use for a wider range of wildlife species applications. Soon it will be reaching the hands of more mainstream adopters or the *early majority*.

From the parts of the story that we can access through the literature, it appears that the use of thermal imaging for wildlife applications began in the late 1960s and into the 1970s. The *innovators* in this field began using the technology to attempt the detection of large mammals in the wild. They primarily targeted deer (Croon 1968; Graves 1972; Parker 1972; Wride 1977), other ungulates such as elk, moose, bison (Wride 1977) and also Polar Bears (Brook 1972). The thermal line scanners they used at the time were nothing like modern thermal-imaging devices used for wildlife today. As we might expect, early thermographic devices were primitive in comparison to their photographic counterparts and were largely unsuitable for wildlife applications due to their low resolution, size, usability and cost. Despite how cumbersome that technology was, and all of the additional paraphernalia required to operate it, these early researchers took to the skies with their thermal-imaging equipment. Using fixed-wing aircraft, they flew over large swathes of habitat to collect their data (Croon 1968; Brook 1972; Graves 1972; Parker 1972; Wride 1977).

Early studies were low in number and patchy in their appearance over time (see Figure 2): there were long intervals between publications in the first two decades of use.

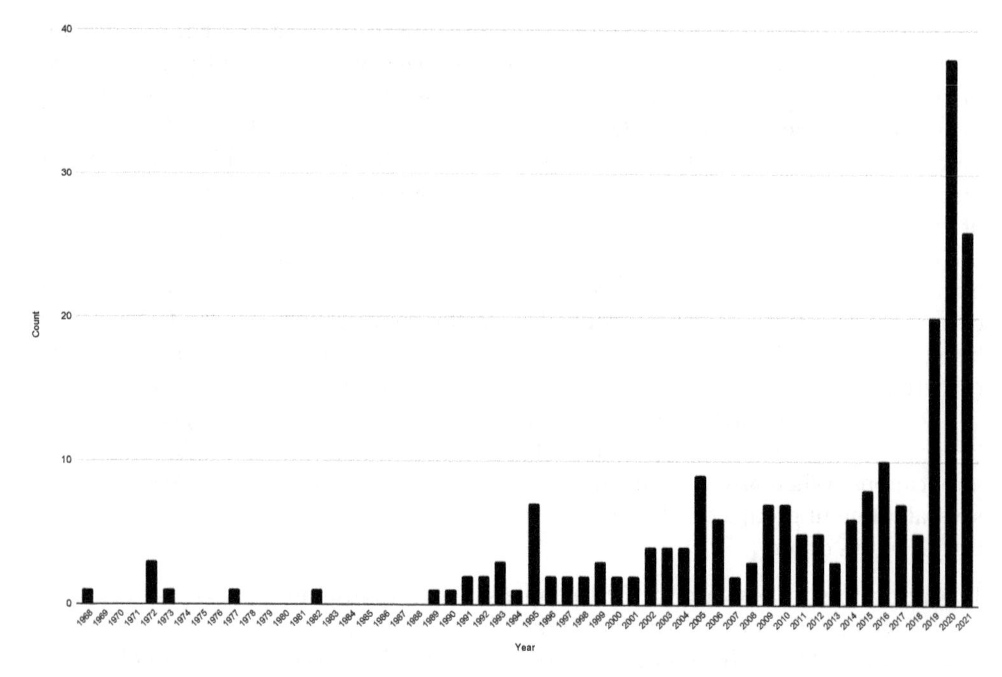

Figure 2 The number of publications that have used thermal imaging for wildlife-related applications from 1968 to 2021 (inclusive).

Kayleigh says:

When I read about the really early days of the use of this technology, I realise how lucky I am with the equipment I have now. I often complain about needing external batteries or extra hard disk drives, but these incredible people were experiencing a whole other level of tech hassle when trying to gather their data!

Some of these projects involved the use of oscilloscopes, VHS tapes and even projecting images onto old-fashioned fabric screens to view their footage. The kit must have weighed tons!

We are so lucky that things have moved on since then, and the technology is becoming increasingly user-friendly and, thankfully, lighter too. This has finally allowed for a rapid expansion in the use of thermal imaging for wildlife in recent years.

Figure 3 Word cloud to show the relative representation of species and species groups in the published literature where thermal imaging was used for wildlife-related studies and projects between 1968 and 2021 (inclusive). The size of each word is proportional to the number of studies on that species or group.

From 1989 onwards, however, papers were released consistently on the subject year on year, though their frequency in number was still low (1–10). It wasn't until 2019 that things started to progress; since then we've seen 20–38 papers produced per year on the subject. Yet the literature seems to massively underrepresent the practical application of the technology in the field where there has been a particularly steep rise in adoption over the last few years.

The relative frequency of papers on each wildlife group or species group varies dramatically and this is illustrated visually in Figure 3. Deer receive considerable attention throughout the history of this field, at least in part due to their economic importance, which drives their intensive management. Deer accounted for a large proportion of the first species group of wild animals to be studied using thermal imaging (Croon 1968), and since then more studies and reports including the use of thermal imaging for deer have been published than for any other group overall (Graves 1972; Parker 1972; Wiggers 1993; Naugle 1996). The second wildlife group that researchers began to study was birds, in the 1980s and 1990s (Best 1982; Boonstra 1993; Sidle 1993; Garner 1995; Liechti 1995; Benshemesh 1996; McCafferty 1998; Fortin 1999). Thermal work on bats started in the 1990s (Kirkwood 1991; Kirkwood 1993; Sabol 1995; Lancaster 1997). A naturally challenging group to study, these fast-flying and mostly nocturnal creatures (Kunz and Parsons 2009) were an obvious target species for this technique. Though the first thermal-imaging study of marine mammals appeared in the 1970s (Brooks 1972), the majority of studies on this group have been generated in the 2000s to 2010s (Armstrup 2004; Willis 2005; Burn 2006; Barbieri 2010). Marsupial studies did not appear until 2000 (Dymond 2000) and, with the exception of a few studies (Grierson 2002; Gonzalez 2016; Hampton 2016), most work in this area has been conducted from 2019 onwards (Lethbridge 2019; Narayan 2019; Corcoran et al. 2020; Hamilton 2020; Vinson 2020; Witt 2020; Baker 2021).

1.3 Advantages of using thermal imaging

Gaining accurate wildlife data can be inherently challenging. The type and levels of these challenges vary depending on each specific wildlife application scenario. Thermal imaging can be used to help tackle some of these challenges while providing a range of benefits, including: minimising disturbance, improved detection, increases in accuracy, better efficiency, reduced costs and improvements to health and safety conditions.

1.3.1 Minimising disturbance

As a passive remote-sensing technology, thermal imaging does not emit any light or sound,[1] making it a valuable tool for non-invasive work with wild animals (Dezecache et al. 2017). Since it can be used in complete darkness, this also means that we can study nocturnal and crepuscular species by conducting night-time surveys without the need for illumination of the animals (Kirkwood and Cartwright 1993). Thermal imaging is not just for night-time use as it can be used in all light levels, so we can use this technology to reduce disturbance when working with a wide range of diurnal species too. Using this technology can help to minimise the potential for disturbance to both target and non-target species, allowing us to not only reduce potential consequences for animal health and welfare but also to reduce the impact of disturbance on our data too.

1.3.2 Improved detection and accuracy

Thermal imaging can be used to help us find animals that are difficult to detect with the human eye (Christie 2016). Wildlife studies and surveys have historically relied on direct observation by humans in the field, yet relying on human sensory systems alone to detect animals is naturally inaccurate because our perceptual capabilities are considerably limited (Bossomaier 2012). These inaccuracies can be worsened by tiredness, making the resulting data even less reliable (Chretien 2016). Given that night work, long hours and fatigue are associated with many wildlife-related roles, we should be mindful of their impact on our

1 With the exception of some cooled devices which can be noisy due to their cooling systems. These are rarely used for wildlife applications, however.

data. Many studies relying on human observation use statistical modelling approaches to correct for errors in sampling (Buonaccorsi and Staudenmayer 2009). However, when we use thermal-imaging technology correctly, we can improve detection rates and reduce errors in the data we collect, which in turn reduces the need for reliance on post-sampling correction techniques.

In visual terms, to successfully detect an animal we must be able to distinguish our target, or the 'signal', from the background habitat, or 'noise'. Several factors, including our limited attention (Cuthill 2019) and subconscious biases based on previous data or information (Joos et al. 2020), can lead to us missing these signals because they are naturally filtered out by our visual sensory systems. Missing a signal from the target animal in this way results in a *false negative result*. The same mechanisms can also result in a *false positive result*, where we report the detection of an animal when none were present. When used appropriately, thermal imaging can achieve high signal-to-noise ratio to help avoid, minimise or reduce *false negative* and *false positive results*. This contributes to improvements in detection accuracy by making it easier to discriminate the target animal from its surrounding habitat. With this in mind, it is unsurprising that researchers have reported that thermal imaging has shown considerable accuracy in comparison to traditional wildlife survey methods, which rely on human observers (Leichti and Bruderer 1995; Perryman et al. 1999; Austin et al. 2016; Lethbridge et al. 2019; Bushaw et al. 2021). Figure 4 shows how much easier it can be to distinguish between animals and their surroundings using thermal imaging versus photographs.

Many wild animals use crypsis to avoid being detected by their potential predators or prey (Edmunds 2004), which can often make it similarly difficult for human observers to find

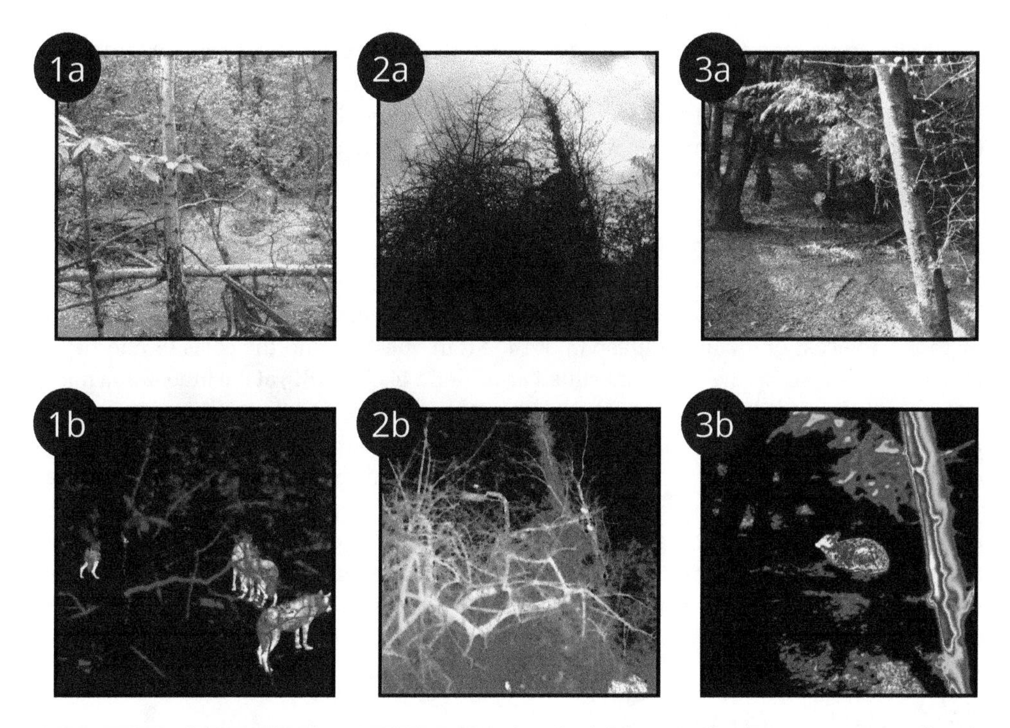

Figure 4 Paired thermal images and photographs of wildlife in the field: 1. European Wolf pack *Canis lupus lupis*: (a) thermal, (b) photograph. 2. House Sparrows *Passer domesticus*: (a) thermal, (b) photograph. 3. Muntjac Deer *Muntiacus reevesi*: (a) thermal, (b) photograph.

and study using traditional visual methods. This can be alleviated by switching to the use of thermal imaging as demonstrated in studies focusing on visually cryptic animals including deer (Chretien 2016), koalas (Corcoran, E. 2021) and owls (Boonstra 1995) to name a few.

1.3.3 Efficiency

Thermal imaging can also help to make wildlife projects much more efficient overall when compared to traditional survey and monitoring techniques. Using thermal imaging to detect kangaroos was much more effective than traditional methods using human aerial observers (Lethbridge et al. 2019): the latter technique being the conventionally accepted method adopted for the detection of kangaroos. These surveys involve aircraft carrying expert observers in challenging conditions at a considerable expense to organisations involved (and also pose a risk to human observers). Lethbridge et al. (2019) found that they detected two-thirds higher population density of their target species when using thermal imaging than the conventional method of human aerial observers. From these results, thermal imaging was more efficient, not only in terms of accuracy but also in terms of cost-effectiveness.

In the UK, surveys of large buildings for bats can often involve large numbers of surveyors. Usually conducted around dusk and dawn, these surveys rely primarily on direct observation and ultrasonic acoustic detectors (Bat Conservation Trust 2016). To increase efficiency of these surveys, thermal imaging is now used to help improve detection rates and to make such surveys more efficient overall, particularly on larger structures where it can considerably reduce the number of surveyors (Fawcett Williams 2021). Use of thermal imaging for bat surveys can streamline the process, reducing time required to plan, execute, analyse and report on them.

Using drone-mounted thermal-imaging cameras can be especially useful in improving efficiency. For example, when Bushaw et al. (2021) used such equipment, they found that not only did they improve their detection rates, but they also completed this work in a third of the time when compared with traditional surveying techniques.

1.3.4 Cost

Equipment can be a significant financial investment, but the returns can be far greater. It is important to consider, as Gill (1997) rightly mentioned in their study of deer censusing methods, the cost of the likely lifespan of the kit in question and the costs of alternative methods. In Gill's case, their imager, valued at around £44,000 (GBP) at the time, was a minor figure in comparison to the usual labour-intensive methods and their subsequent costs.

Direct observation by humans can be a relatively inexpensive method of wildlife detection when using unpaid or low paid surveyors (an unfortunately common practice), while using highly skilled and fairly paid professionals can be relatively expensive. All of these surveyors, however, are human and are subject to errors and biases inherent to their sensory systems. Human error can be far more expensive than the costs of thermal imaging, so it is always worth seriously considering this when weighing up whether or not a project can afford it.

1.3.5 Health and safety

Using any technology that can reduce manual survey effort has the potential to improve working conditions for wildlife professionals. A lot of thermal-imaging systems allow us to sample a much greater area by improving our detection distance (range) and/or broadening our functional width of view. Where this is the case, we can reduce person hours required for the task – this might mean we use less people on the ground, as is often possible for bat survey applications, or that individual personnel can cover more ground in less time. Either way, this can potentially reduce (often heavy) workloads for individuals or teams of wildlife professionals with implications for their health and well-being. Arguably, this can bring the highest level of benefit to nocturnal surveyors because night work can have serious negative consequences for both health and safety. The long-term effects of shift and night work have been well documented, and they pose an increased risk of the onset of a range of chronic and fatal conditions (Ramin et al. 2015). From a safety perspective, night work can lead to fatigue which puts staff at a higher risk of accidental injury, or worse (Williamson et al. 2011). Reducing the load of this type of work on surveyors, could therefore have huge benefits in both the short and the long-term.

1.4 Challenges and limitations of thermal imaging

While thermal imaging can be an excellent technique to help us to collect better data for wildlife-related applications, it is important to also consider the challenges associated with it and its limitations. Thermal imaging is not suitable for all wildlife applications and it is important to consider each potential application on a case-by-case basis. However, when these limitations are taken into account and the technique is applied appropriately, thermal imaging can be a very powerful tool in our wildlife toolkit.

It is vital to be clear on what thermal imaging *cannot* do. Before we discuss unsuitable applications, we must first consider why this issue is so critical: the implications of misuse of the technology are potentially devastating. This is particularly precarious in situations where the consequences of *false negatives* (where an animal is present but not detected) could later result in disturbance, destruction of habitat, injury or death of individuals, groups or even entire populations of wildlife species. For species on the edge of extinction, this is not a technique to be used lightly. This is clearly demonstrated by Smith et al. (2020) in their assessment of its efficacy when used for the detection of Polar Bear dens in Alaska (see Chapter 6 for more on this study). It is important to note that thermal imaging cannot allow us to see through solid objects (or water). It can allow us to see through gaps in vegetation, but a clear line of sight is required as infrared waves cannot penetrate through solid objects. This may seem obvious, but there is a common misconception that it is possible to detect animals within or behind solid objects – such as bats within a tree roost or bees inside a chimney-stack – which it cannot. We may see patterns or signatures that suggest animal presence, but it is important to be clear that what we are seeing is only from the surface of the object and does not allow us to view what lies underneath.

For applications that have been already demonstrated to be suitable, and for those likely to have good potential, success is still limited by several factors including expertise, equipment and environmental conditions. It is vital to ensure that these are appropriate for the specific task at hand. First, it is imperative that wildlife professionals using thermal imaging have the relevant expertise to be able to apply it effectively (see Section 3.5 for further information on expertise requirements). Second, they must then select equipment with the appropriate specifications to meet the needs of their specific wildlife application (see Chapter 4). Third, sampling should be carried out under environmental conditions that allow good detectability (see Section 2.2 for more on this).

2. Foundations

As we learned in the previous chapter, thermal-imaging devices operate by detecting infrared waves and transforming them into visual data. Using this technology can bring a range of benefits to wildlife projects. However, in order to be able to collect and use this kind of data effectively, it is important to understand some of the foundational concepts that define what we can 'see' using this specific technology.

It is important to note that the theory we cover here is not just for academic purposes: it has very real and practical implications for detectability of wildlife, accuracy of results and ultimately, success or failure. Knowledge of these principles help us to design and implement better thermal-imaging studies or projects, allowing us to efficiently obtain usable, meaningful wildlife data when using this technology in our work.

2.1 Setting the thermal scene

Before we discuss the detectability of wildlife, we first need to set the thermal scene to understand what it is we are looking at when we are using thermal imaging. To do this, we will explore some of the underlying physical processes and properties involved, as well as the key terminology used.

2.1.1 Underlying processes

The thermal world around us is in constant motion with many different physical processes going on around us all the time – it never stays still. While a thermal image may catch a snapshot, this represents a point in time where many dynamic processes are ongoing. This is something that we should be mindful of when using thermal imaging: we are always viewing an image that represents multiple dynamic processes in motion.

The processes and concepts introduced in this section (below) are governed by a number of fundamental laws, including: Kirchhoff's Laws, Stefan–Boltzmann Law and Planck's Radiation Law. If you would like to learn more about these laws and examine the equations used to express them in the context of wild animals, you can find excellent accounts in Havens and Sharp (2016).

Heat-transfer processes

Thermal imaging relies on the detection of infrared waves. The mechanisms for the transfer of these infrared waves, or heat, affect the thermal images we collect and their usefulness for wildlife applications. It is important to consider such heat-transfer mechanisms when we are planning, conducting and analysing data for thermal-imaging projects.

These heat-transfer mechanisms are:

- conduction
- convection
- condensation/evaporation
- radiation

Realistically, the majority of what we see with a thermal-imaging device is due to radiation. In the context of wildlife, however, it is also common to see the effects of convective cooling of surfaces, such as leaves in the breeze or the wings of bats in flight. Occasionally we will also see the effects of conduction, such as thermal footprints left by animals walking on a surface. If we are very lucky, we might even see the evaporation of sweat from animals in the field.

Radiative heat transfer

As radiation is responsible for most of what we see in a thermal image, we need to dig a little deeper into how it is transferred. Radiative heat transfer occurs by the following four processes:

- emission
- absorption
- reflection
- transmission

Infrared radiation is not only being emitted by all objects in the environment; it is also being absorbed, reflected and transmitted. When we are attempting to detect an animal, we must keep in mind that while it is emitting radiation from its body surface, a profusion of other transfers are taking place within and around it. Figure 5 illustrates these processes.

2.1.2 Temperature

It is a common misconception that thermal cameras display the temperature of objects. In practice, there are actually a number of factors we need to take into account to reach something close to the absolute or 'true' surface temperature. These include:

- apparent surface temperature
- emissivity
- distance
- air temperature
- humidity

Wildlife applications rarely require the calculation or estimation of true or near-true surface temperatures. The measurement or calculation of this parameter (also known as the absolute temperature) is usually only involved in research applications related to thermal biology or thermoregulatory processes (see Section 5.5). This procedure is not necessary for the majority of applications such as those for detection, classification and recognition of wildlife. It is, however, very useful for anyone who uses thermal-imaging

Figure 5 Paired photographs and thermal images of a Scottish Wildcat *Felis silvestris silvestris* taken in bright sunlight. Conditions like this can be challenging due to solar loading and solar reflection.

technology to be aware of what is involved, in order to better understand the underlying concepts and their wider implications.

Some thermal-imaging devices allow us to calculate this temperature within the device and/or within specialist software packages during image processing. When collecting thermal images to be used for temperature measurement, it is important that certain conditions are met and that the operator has the necessary skills to collect these effectively (see Section 3.5). Professional thermography training courses such as the Level 1 Thermography Certification are a good starting point to learn how to carry out this procedure effectively (see the Supplier Directory on page 127 for information on course providers). Equations for the calculation of absolute surface temperature can be found in Öhman (2014) and in Level 1 & 2 Thermography Course Manuals (Infrared Training Center 2010 and 2016).

When temperature measurement is implemented, it is vital that researchers report their methods correctly. Unfortunately, biologists in general are not great at this (Harrap et al. 2018) with almost half failing to report the critical values used. This may be due to incidental omission or, more worryingly, because their calculations are not accounting for key factors. Either way, without them it is impossible for readers to understand how accurate reported temperatures are.

Apparent surface temperature

Apparent surface temperature is an uncorrected temperature value displayed on the thermal camera or in specialist software when viewing images externally to the device. It sometimes helps to consider this value as the temperature that the camera 'thinks' it can see based on the data available to it. We can then provide the additional information that is needed to adjust this value to one that is closer to the absolute surface temperature.

Emissivity

Emissivity defines the relative ability of a surface to emit radiation. Emissivity is an important concept in thermography because it has a huge bearing on how objects appear in thermal imagery: it can make objects look hotter or colder than they truly are.

Emissivity is expressed as a value between 0.0 and 1.0. Theoretically speaking, an object with an emissivity of 0.0 would be unable to radiate at all whereas an object with an emissivity of 1.0 would be considered a perfect emitter (though neither of these exist in reality). The emissivity value of an object is dependent on a number of other factors, including what material it is composed of, the geometry of the object and the angle from which it is viewed. Emissivity values are generated under laboratory conditions and these can be found for many different materials in lookup tables (for examples, see iRed 2021, Öhman 2014, Infrared Training Centre 2010).

Biological surfaces are generally assumed to have relatively high emissivity – usually using values around or greater than 0.95 – but such assumptions can lead to error (Playà-Montmany and Tattersall 2021) and their validity has been disputed (McGowan et al. 2018; Graveley 2020). Emissivity values for skin, fur, feather, etc. are generated using dead specimens because the process involves heating of samples, which could not be carried out on live animals. Even without the inaccuracies that could result from the use of dead specimens, biological surfaces are, by their very nature, highly variable between and within individuals. Therefore, if using them inferentially for temperature measurements in live animals, especially if they are studied in the field, researchers should be mindful of the potential impacts this may have on the accuracy of temperature measurements they may make.

Distance, atmospheric temperature and humidity

The distance between the target object and the thermal-imaging device, as well as the atmospheric temperature and humidity, all affect the apparent temperature reading. These must be corrected for by entering them into thermal camera software or by including them in any calculation to approach an absolute temperature reading.

2.2 Detectability

The majority of wildlife applications require the detection of the body of the target animal species. This is possible with thermal imaging when there is a sufficient difference between the apparent surface temperature of the animal and the apparent surface temperature of its surroundings. This difference – sometimes referred to as **ΔT** – makes it possible to distinguish the signal (animal) from the background noise or clutter (habitat).

The detectability of an animal can be affected by a range of factors, including:

- body size
- body-surface properties (skin, fur or feather coat)
- habitat (type, complexity and physical properties)
- environmental conditions
- behaviour
- viewing angle
- device specifications (see Chapter 4)
- expertise of operator (see Section 3.5 for more on expertise requirements)

These all have an impact on whether or not we can discriminate an animal of our target species from its surrounding habitat.

2.2.1 Body size

The larger the available surface area of an animal's body is, the easier it is to detect (where other factors remain constant). For example, a tiny Pipistrelle Bat *Pipistrellus pipistrellus* in flight can be detected by the FLIR T1030sc (FLIR Systems, Inc., Wilsonville, OR, USA) with a 12° lens at a distance of 392 m, whereas a much larger mammal, such as a Badger *Meles meles*, can be detected at 1,466 m with the same equipment.

2.2.2 Body-surface properties

Properties of the body surfaces of animals, including emissivity, geometry and texture, vary between and within species as well as within each individual animal. These surfaces may predominantly be skin or scales, or they may be fur or feather coats. The properties of the outer surfaces or coats of animals can also vary over time: they may have seasonal cycles (e.g. shedding or moulting) or they may change progressively over the lifetime of the animal (e.g. hardening or thickening). These differences in physical characteristics affect how an animal appears in a thermal image, often with practical implications for detection (see Figure 6).

Certain discrete areas on the surface of animal bodies can appear warmer and therefore more distinct in a thermal image. Some of these are natural 'hotspots' that happen to appear warmer than other areas of the body due to natural lack of insulation or differences in surface properties, such as eye or ear regions (see Figure 7). Others do not just happen to appear warmer: they are thought to be functional areas of importance for thermoregulation and are sometimes referred to as 'thermal windows' (not to be confused with *IR windows* which are hardware described in Chapter 4). These thermal windows include the ears of African Elephant *Loxodonta africana* (Phillips and Heath 1992) and bills of Toco Toucan *Ramphastos toco* (Tattersall, Andrade and Abe 2009) which were identified with the help of thermal-imaging technology (for more details on the discovery and functional importance of these areas see Section 5.5 on thermoregulation and thermal biology). Klir and Heath

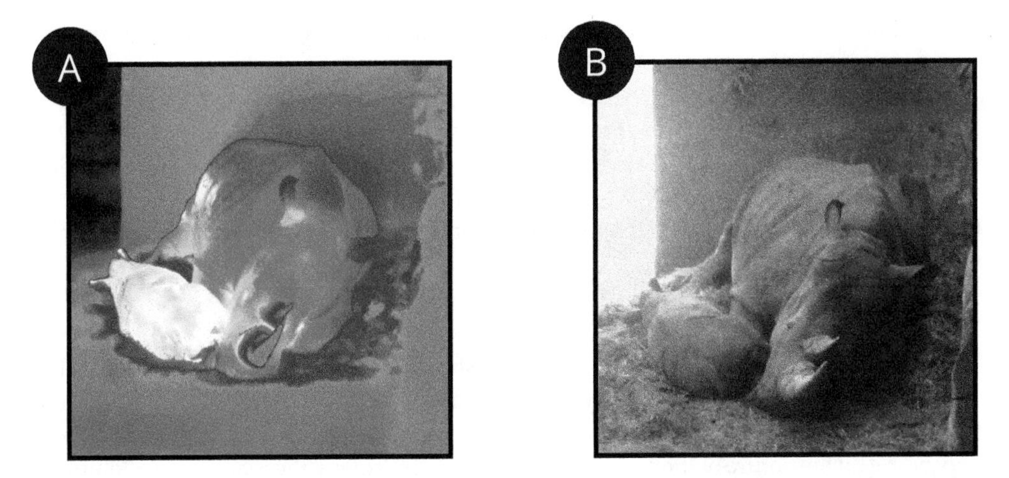

Figure 6 A thermal image of a rhino mother and calf. Notice the difference in the appearance of their body surface: this could be due to a difference in their absolute surface temperatures, but is likely affected by the emissivity and surface texture of the skin which changes as they age.

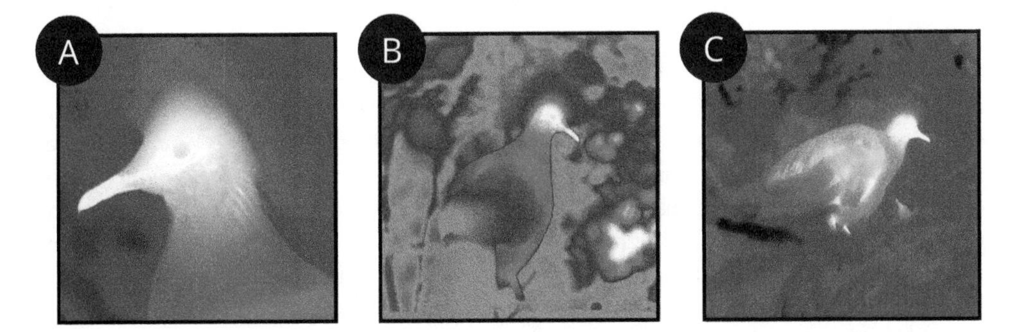

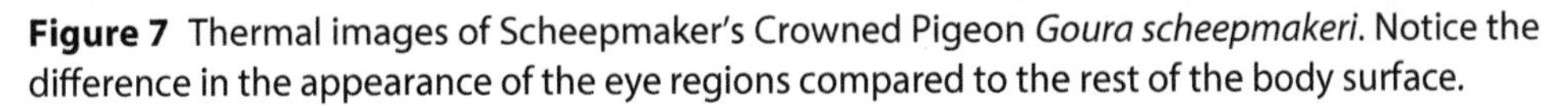

Figure 7 Thermal images of Scheepmaker's Crowned Pigeon *Goura scheepmakeri*. Notice the difference in the appearance of the eye regions compared to the rest of the body surface.

(1992) also found several such areas in foxes using thermal imaging. Using an Inframetrics Model 525 (Inframetrics Inc., Billerica, MA) thermal-imaging camera, they identified several distinct thermoregulatory areas in Red Fox *Vulpes vulpes*, Arctic Fox *Alopex lagopus* and Kit Fox *Vulpes macrotis*. These included parts of the face, nose, ears and lower legs, which had less hair coverage and appeared warmer than the rest of the body surface. These areas, depending on their size, variability and conspicuousness, may aid in detection: particularly useful for detecting well-insulated animals which might otherwise be a challenging task.

2.2.3 Habitat

The surrounding habitat type and complexity can have a big impact on our ability to find wildlife using thermal imaging. It is essential to have a clear line of sight between the thermal-imaging device and the animal, as anything that breaks this can impede detection. Dense forest canopy and tall ground vegetation, for example, can obscure the target. However, when partial sightings are possible through gaps in vegetation, detection can be much easier than using traditional visual techniques due to the thermal contrast between the animal and its surroundings.

Some non-target objects in the scene can cause *false positives*, and this varies between habitats. The extent to which this is an issue will depend on the physical properties of the non-target objects and their relative similarity in size and shape to the target. In reality, these false positives are usually weeded out during analysis, but if they occur regularly, they can be problematic due to the time and effort required to check and exclude them.

2.2.4 Environmental conditions

To detect wildlife using thermal imaging, we should aim to sample in conditions that maximise the potential difference between the apparent background temperature and the target animal. Finding the optimal conditions for sampling can vary in difficulty depending on the nature of the application. However, in general there are some established environmental factors that should be considered, including:

- sunlight (solar loading and reflection)
- thermal cross-over
- humidity
- wind

Some of these factors may be more or less important depending on your target species and method of deployment. Different applications can also require different environmental conditions for optimal detection. For example, Leichti and Bruderer (1995) found that clear skies during nocturnal sampling for birds allowed for better detection than cloudy ones. Conversely, Witczuk (2018) found that clear skies yielded the poorest results when seeking primates during the day. Testing prior to sampling helps to establish suitable conditions and is especially important for novel species applications or situations. Such results can inform good planning to ensure that sampling is timed to be carried out in periods that are most likely to provide optimal conditions (these topics are covered in detail in Chapter 3).

Sunlight

Periods of bright sunlight can cause issues due to solar loading, reflection and even potential damage to equipment (which can occur if exposed to direct sunlight).

In order for us to be able to detect our target animal, there must be a large enough ΔT: the difference between the apparent surface temperature of the animal and the apparent surface temperature of its surroundings. Solar loading and solar reflection of objects in the environment can cause significant issues, which have caused challenges for researchers attempting to detect animals during the day (e.g. Kays et al. 2019). In some cases, these issues are unavoidable and could make thermal imaging an unsuitable tool where such conditions make distinguishing the target animal from the background habitat difficult or even impossible. More often, however, these issues can be avoided with careful forethought and planning by professionals with appropriate expertise. Timing is key (we will discuss this in detail later in Chapter 3).

In order to forecast optimal timing for drone-mounted thermal imaging of wildlife, Burke et al. (2019) used land surface temperature (LST) data generated by satellite. This appears to be a very useful tool for planning the timing of sampling.

Thermal cross-over

Periods of thermal cross-over or inversion occur within daily cycles. When this happens, objects become indistinct in thermal images due to low thermal contrast (Felton 2010). These periods can come about at any time during the daily cycle, so it is important to find out when they usually fall in or around the location of interest. For daytime applications, testing to ascertain when these periods fall can be invaluable.

Humidity

The amount of water molecules in the atmosphere affects the transmission of infrared waves (Infrared Training Center, 2010). Humidity can have effects on image quality, but the relative effects of humidity on both temperature and detection are quite low. More importantly, it is important to ensure that devices have the appropriate level of ingress protection if they are to be used in high humidity environments (see 4.1.8. for more information on ingress protection).

Wind

Air movements can cause changes in apparent surface temperature and physical movements of objects in the thermal scene (e.g. a breeze causing convective cooling and movement of leaves on trees). In many cases, the effects of wind will be trivial, but it can cause issues for automated analysis procedures.

2.2.5 Behaviour

Earlier we discussed the physical aspects of our target animal, but their behaviour is just as important to consider. Some behaviours may inherently make animals more conspicuous than usual. Other behaviours may mean that animals are easier to detect because they aggregate in discrete areas or in larger numbers. Some researchers have chosen to wait until animals are asleep or are in sleeping areas to make detection easier (Jumail 2020).

Activity may affect detectability due to movement itself and because of thermal phenomena associated with activity. Movement can increase an animal's conspicuousness, which can be especially useful to aid detection when the difference between the animal and their habitat are low. From a sensory perspective, movement provides us with strong indications to help us identify an object (Bossomaier 2012). This can help us considerably when we are manually analysing footage or viewing live in the field. From a thermal perspective, activity has the potential to change the apparent surface temperature of an animal. The extent to which this manifests likely varies dramatically between different species. Some animals display differences in apparent surface temperature during daily cycles (e.g. Hedgehogs: see South et al. 2020). During periods of inactivity, some species may lower their body temperatures in order to conserve energy. This can vary in length, from short bouts of torpor to long seasons of hibernation. This usually means that such animals blend into their environment in a thermal image, making them undetectable.

3. Methods

Well-thought-out methods are key to the efficient and productive application of thermal imaging for wildlife-related projects or studies. In this chapter, we will consider each of the following topics that can help us to achieve this:

- operations workflow
- levels of use
- deployment methods
- combining techniques
- expertise requirements

3.1 Operations workflow

An operations workflow can be used to describe a thermal-imaging procedure throughout the entire process. Having an operations workflow helps to give structure and smooth the process from concept to completion. It can also help us to avoid unnecessary, and often costly mistakes along the way. It may require an initial investment of time to set up, but will save considerable time and effort overall.

The stages are:

- screening
- scoping
- planning
- pilot or testing
- sampling
- analysis
- reporting
- review

These stages are common to many thermal-imaging applications, including those involving the detection, classification and counting of target wildlife species. As each application is unique, some projects will require all stages and others will not. However, if you are new to the subject or establishing a workflow for a novel application, this process can be used as a starting point. For novel or complex applications, an iterative process may be required.

Many projects fail because they do not have an established workflow process or skip key stages. A common example is where thermal-imaging equipment is deployed for the first time during the sampling phase, without a Pilot or Testing Stage being carried out. This can lead to a range of issues that can have further knock-on effects later in the project or study and beyond.

3.1.1 Screening stage

During the screening stage, we begin to assess whether thermal imaging is likely to be appropriate for the aims of our project and the species involved. While thermal imaging is an attractive and powerful tool for many wildlife applications, it is not suitable for all species and situations. Therefore, it is important to carefully consider the suitability of this technology for the work in question.

We can begin to evaluate the suitability of this technology by researching existing work that has applied thermal imaging for our target species and situation. It is important to review the methods critically in order to examine the use and relevance of previous studies relative to our current work. For novel applications where there is no prior information, we can examine previous studies that have focused on species of similar size and type. In either case, we should evaluate the similarities and differences in target and non-target species, habitat and environmental conditions between those studied and those we plan to study or survey.

Where there is an absence of available information, we can calculate or estimate detection distances for the target species based on body size and equipment specifications. While such numbers may be purely theoretical, they will give us a good indication of how useful this technology is likely to be.

When screening for a novel application, there are a number of important considerations in relation to the target and non-target species and their surrounding environment. These parameters can impact the suitability of thermal imaging for the application. Figures 8a and 8b illustrate some key factors to consider during the screening phase.

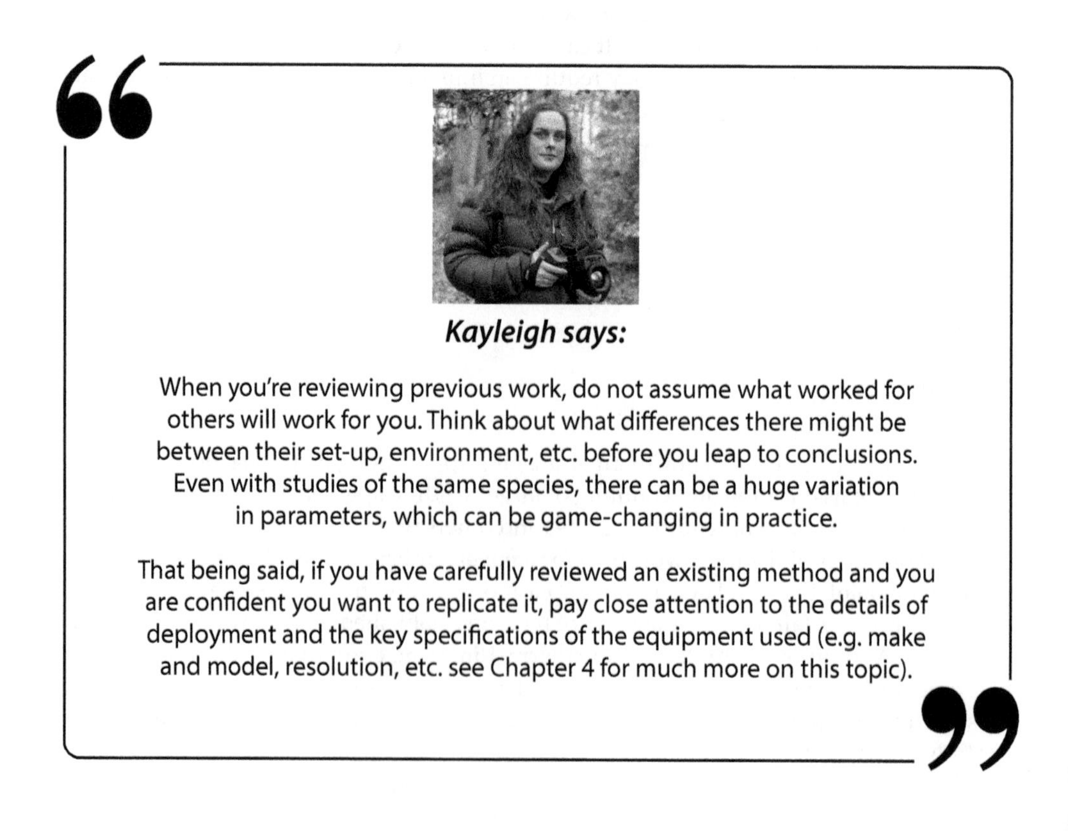

Kayleigh says:

When you're reviewing previous work, do not assume what worked for others will work for you. Think about what differences there might be between their set-up, environment, etc. before you leap to conclusions. Even with studies of the same species, there can be a huge variation in parameters, which can be game-changing in practice.

That being said, if you have carefully reviewed an existing method and you are confident you want to replicate it, pay close attention to the details of deployment and the key specifications of the equipment used (e.g. make and model, resolution, etc. see Chapter 4 for much more on this topic).

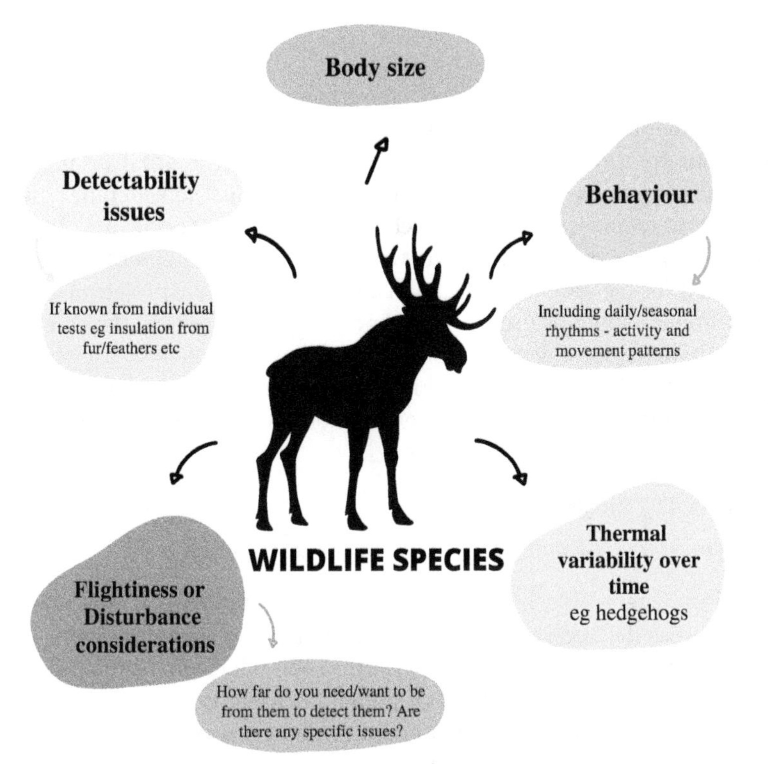

Figure 8a Factors to consider during the screening phase: wildlife species (target and non-target).

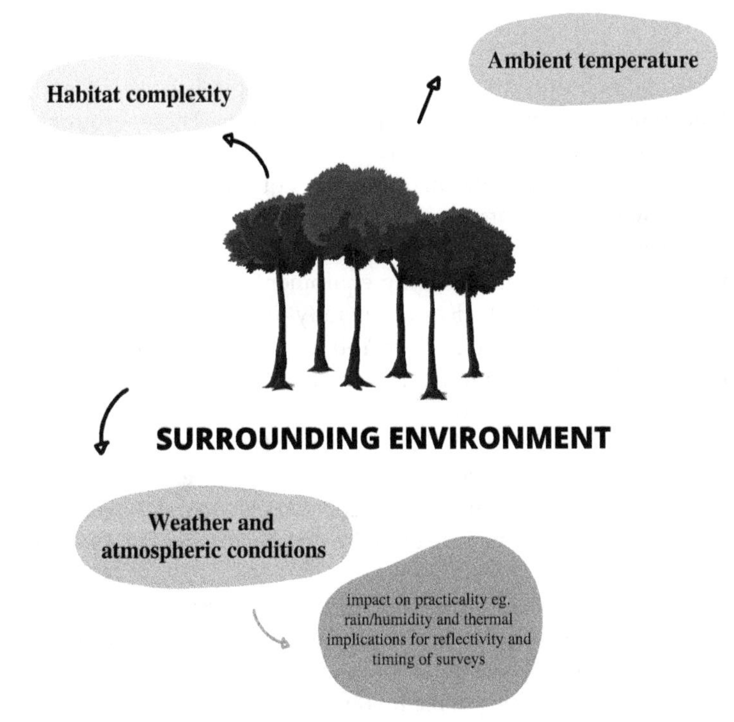

Figure 8b Factors to consider during the screening phase: surrounding environment.

3.1.2 Scoping stage

Once we establish that thermal imaging has the potential to be a good fit for our application, we move on to the scoping stage. Here we will further evaluate this technology and how well it suits our specific application. We will also consider alternative approaches and technologies to compare their merits and limitations. We must consider the type and scope of the data output requirements of each technology, as well as species and habitat-specific characteristics. By completing this stage, we can determine whether thermal imaging is the best method for the specific application and the aims of the project or study, or if an alternate technique should be used. Havens and Sharp (2016) provide an excellent account of alternative remote-sensing technologies, including radar, sonar, trip (trail) cameras, image intensifiers and radio telemetry. It is also possible that you may need to use thermal imaging in combination with one or more non-thermal techniques (see Section 3.4).

3.1.3 Planning stage

'If You Fail to Plan, You Are Planning to Fail' – Benjamin Franklin

At the planning stage, we define the detailed objectives of the thermal-imaging project or study and organise the necessary resources and logistics to achieve them.

It is important to clarify objectives such as the central hypothesis (if applicable), types of data output and the required level of detail. Defining these at the outset will help streamline processes and avoid potential issues at later stages of the workflow. This will also help to identify appropriate methods, procedures and resources specific to these end-goal requirements.

Resources

Once we have a defined output, our next step is to begin organising and scheduling resources to facilitate the completion of the project or study. These resources can include equipment, time, expertise and permissions.

One central resource is the thermal-imaging equipment itself. First, we need to define which thermal-imaging device or devices you need for your specific application and data objectives (see more on equipment in Chapter 4). If you do not already have access to the appropriate thermal-imaging equipment, you will need to organise hire or purchase in advance. Some devices are readily available 'off the shelf' and others require long lead-times, so planning is key. You may need to consult a local supplier to facilitate shipping and delivery. Some devices and components are subject to government-controlled restrictions on import and export (Kays et al. 2019; Semel 2020), so if you plan to travel across political borders with this equipment for fieldwork, make sure that you can do so legally. If you have any doubts, contact your supplier or the manufacturer for advice on any specific restrictions on the device(s) you plan to use. As well as the thermal-imaging device(s), you may also need to organise the procurement or distribution of other equipment and accessories.

At this point, we must also define the human resources required. This includes planning who will be responsible for the deployment, operation, analysis and reporting of the thermal-imaging data. Sometimes all tasks are carried out by one person, but in most cases, there will be multiple people involved in one or more tasks in the process. If handovers are required, it is important that these are considered and planned in advance to avoid unnecessary delays or errors. It is also crucial to ascertain what skills or expertise (see Section 3.5) are required for each thermal-imaging task to ensure that the individual or team is appropriately trained to carry out their task(s) effectively.

Timing

It is also vital to plan the timing of data collection or sampling (surveys or trials) for our target species and the technology we plan to use. We can refer to the existing literature for survey timing windows, periods of activity or behaviours of interest. Timing can be crucial for the use of thermal-imaging technology, where particular times of day and/or season may provide optimal thermal conditions that allow good thermal contrast between the target species and the background of the surrounding habitat. We can also use the next (pilot or testing) stage to help us determine the optimal thermal timing for sampling. It is highly desirable to find an overlap between species and thermal requirements, where possible, to maximise the probability of detection.

Heat emission, absorption and reflection from both target species and their surrounding habitats vary throughout the course of the day (and night). It is important to select optimal windows of time during any 24-hour period that are most likely to give maximum thermal contrast between the target species and the background habitat. Conversely, it is also important to avoid periods of thermal cross-over or inversion which occur at certain times of the day (Felton et al. 2010). The timing of such cross-overs can vary according to the thermal properties of the areas in question, as well as the conditions in the surrounding environment. When they occur, the contrast between the target species and the background habitat will be low, making them indistinguishable from one another. This is particularly troublesome when these periods coincide with the peak activity times of the target species. In these cases, it may be necessary to re-evaluate thermal methods or seek alternative approaches. It has been suggested that polarimetric thermal-imaging techniques could help improve the discrimination of target objects from the background for military and security applications, but it does not appear to have been utilised for wildlife detection applications to date (Fust and Loos 2020).

Several authors have reported that time of day has had an impact on their images and/or their ability to detect their target species (including Mulero-Pazmany 2014; Witczuk et al. 2018; Rahman and Rahman 2020; Bushaw et al. 2021). For example, in their comparative study on thermal imaging versus traditional survey methods, Bushaw et al. (2021) found that their morning sessions allowed better conditions for the detection of ducks than in the afternoons. This may have occurred for two main reasons: the thermal conditions and the behaviour of their target species. Morning surveys may have been cooler, offering good thermal contrast, whereas afternoon surveys may have been warmer, allowing little thermal contrast between the ducks and their surroundings. Additionally, from a behaviour perspective, the ducks may have been easier to detect because they tended to be more active in the morning. Selecting an appropriate time of day to collect thermal-imaging data can be critical to success, so it is vital that this is considered carefully when planning your project.

Non-target species

When planning, it is also important to consider not only the target species but also the known or potential presence of other non-target species. Other species in the area are a possible source of error (Croon et al. 1968). However, there are often ways to minimise the potential for confusion between target and non-target species. These approaches can include:

- Collecting reference images of the target and non-target species for operators/analysts to compare in the field or during later analysis. Multiple reference images should ideally be taken from a range of distances, angles and conditions with the same equipment used during sampling.
- Providing training on the identification of target and non-target species to operators/analysts.
- Altering timing to avoid periods when non-target species are most likely to be present with the target species (although of course, this is not always possible).

3.1.4 Pilot or testing stage

Where they are used, a pilot or testing stage can be extremely helpful and may often even be critical for some wildlife applications. This stage should always be included for novel species applications. It is also important to complete this stage when using new types of equipment or method. Conducting this phase can help us to identify any issues specific to the species involved, systems or their set-up. These issues can then be addressed prior to the start of data collection during the sampling stage. This may vary in time and complexity: it may simply take the form of a short test run or it may be a longer, more comprehensive pilot or feasibility study.

This stage is the perfect opportunity to test or 'ground truth' our expectations of the capabilities of the technology before we go out to collect data. This can be done with free-ranging animals in the wild or under more controlled conditions using captive individuals. Testing with free-ranging animals in the wild allows a more realistic appreciation of what the technology can achieve. It also previews potential habitat challenges in terms of the application of the device from a thermographic perspective, and from a practical standpoint in terms of logistics of accessing and deploying the equipment. In their study on the effectiveness of bat deterrent devices, Schirmacher (2020) used ground-based thermal-imaging cameras to monitor the activity of free-flying bats under different experimental conditions. They conducted a feasibility study prior to sampling for their subsequent comparative study. This allowed them to better understand the process of using the technology, which gave them the opportunity to improve aspects at every stage in their workflow. In their comparative study, they report having had only very minor issues that had a minimal impact on their results. It is likely that this is because they had already ironed out the main issues from what they learned in their feasibility study. In another study, Jumail (2020) carried out testing to optimise their planned methods for surveying of primates using a thermal-imaging system on board a boat. This allowed them to determine optimum conditions for their surveys, including the timing, weather and boat speed.

However, initial testing in the field may be challenging, time-consuming or simply unrealistic. For species that are rare, elusive, sensitive to disturbance or only found in remote or challenging environments, testing in the wild may require considerable time and effort to find the target species. In such cases, it may be more sensible to use captive individuals for the purposes of initial testing. In preparation for the sampling stage of their study on wild rabbits, Psiroukis et al. (2021) used a testing phase to determine optimum flight parameters for a thermal drone. Using the FLIR Vue Pro, they conducted night flights

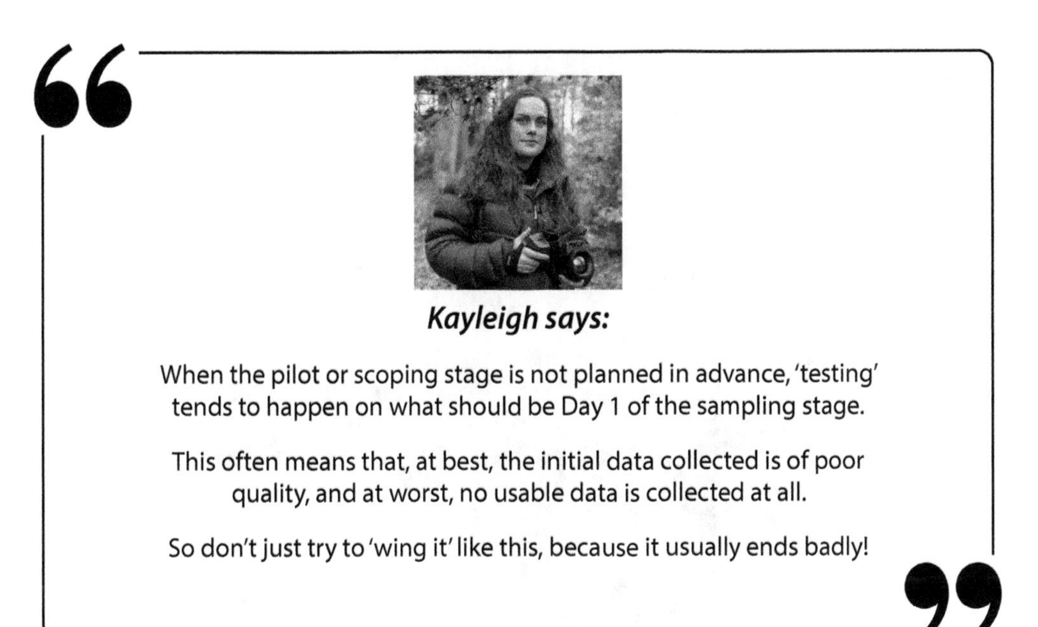

over captive rabbits, which allowed them to design a flight plan based on the findings of their test. This allowed them not only to be able to gain sufficient image quality to distinguish the target animal from the background habitat but also to minimise disturbance to the animals themselves.

Whether collected in the wild or under more controlled conditions, images or videos from testing can be used to assess the quality of the imagery and its likely usefulness for our application. This can have a huge impact on later stages and may provide critical information for the analysis stage, especially where automated or semi-automated analysis procedures are required. We can also take this opportunity to collect reference images, which can then be used to assist with classification or identification (see 3.1.6 Analysis stage). Detection distances (as calculated or estimated during the planning phase, see Section 3.1.3) can also be verified or adjusted by taking sample images from known distances from the target animals.

3.1.5 Sampling stage

In some disciplines, we might call this the survey stage, the active monitoring, or, in research applications, trials or experiments. Regardless of the terminology, the sampling stage is where we actually collect the main body of data for our project or study. This stage often involves the use of tripod-mounted cameras such as the set-up shown in Figure 9, though there are many other options available (see Section 3.3 Deployment Methods on page 32).

This stage should be scheduled at the planning stage, according to the specific aims and requirements of the project or study and the target species involved. It should allow a sufficient amount of time to collect the required data with the provision of sufficient contingency to allow for common challenges such as inclement weather and logistical delays.

Once data is collected, it should then be organised, stored and backed up in preparation for the analysis stage.

Figure 9 Sampling using a tripod-mounted FLIR T1030sc thermal-imaging camera.

3.1.6 Analysis stage

After thermal-imaging data is collected, the data must be processed and/or analysed to extract data outputs with maximum accuracy and reliability. The steps required in this stage will be determined by the type(s) of data output that was defined at the planning stage (see Section 3.1.3).

Post-processing

When using radiometric data, still thermal images and thermal video can be processed using specialist software. This includes the process of image optimisation (thermal tuning). An experienced thermographer can refine raw data into processed images or videos that maximise the potential for successful detection of target species and the extraction of other, more detailed information where required. The goal is to maximise the contrast between the target animal and the background habitat. If non-radiometric data is used, the resulting still or video files will be of a standard format (e.g. JPEG, AVI), which cannot be optimised or tuned in thermal software packages. Figure 10 shows the difference post-processing can make to the appearance of thermal images.

During post-processing, it may be desirable to crop images or videos. Both radiometric and non-radiometric images can be cropped spatially to include/exclude particular areas of the images. Radiometric and non-radiometric video files can also be cropped or snipped temporally to select particular timescales, video file lengths or durations to allow for easier data handling and future analysis.

At this sub-stage, you may also need to complete file conversions. For example, you might record a particular file type (as dictated by your chosen device) but your analysis programme(s) may require the input of a different file type. It is important to ensure that

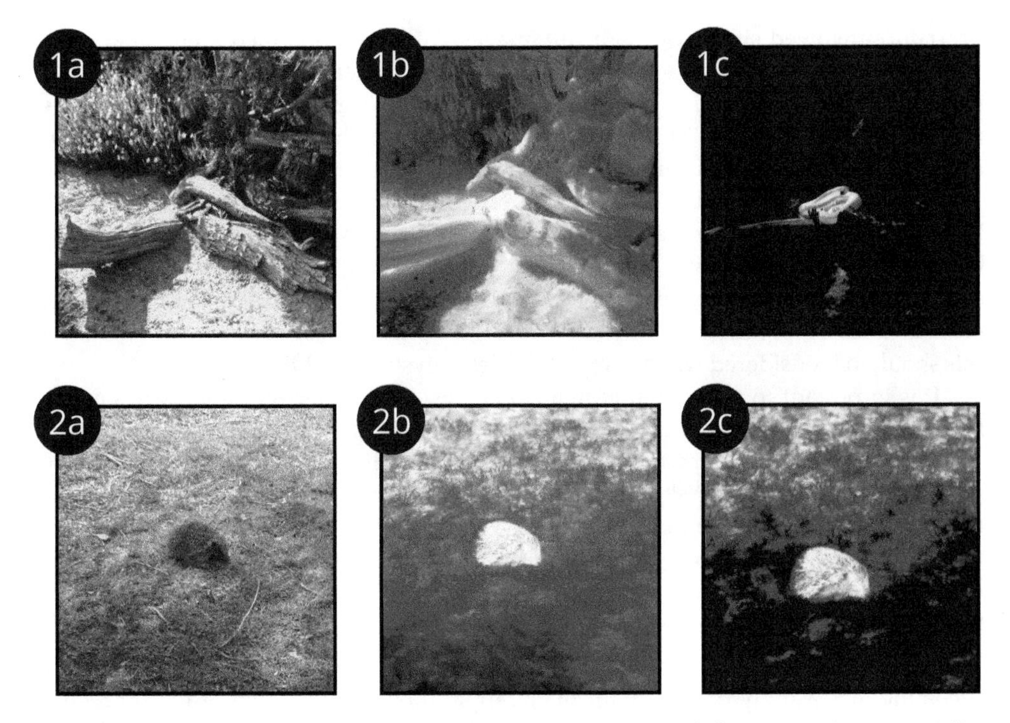

Figure 10 Photographs and radiometric thermograms before and after processing in specialist thermal software. 1. Adder *Vipera berus*: (a) photograph, (b) unprocessed radiometric thermogram and (c) radiometric thermogram after processing in specialist thermal software. 2. European Hedgehog *Erinaceus europaeus*: (a) photograph, (b) unprocessed radiometric thermogram and (c) radiometric thermogram after processing in specialist thermal software. Notice the difference in contrast between the target animal and the surrounding habitat across the different image types.

any conversion you make can be done without compromising your data integrity. Always make a backup copy of the original file prior to conversion. This reduces the potential for data loss and allows you to keep the file in its original format for long-term storage or until you are sure you no longer need it (as appropriate).

Manual analysis

Manual analysis can be useful for some applications. There are many situations where simply watching thermal images directly in real-time on a device or external screen in the field is not sufficient to detect wildlife. Doing so risks animals being missed.

When searching for Polar Bears using airborne thermal-imaging cameras, Brooks (1972) found thermal trails left by the Polar Bears in the snow during the analysis stage. After processing and manually reviewing their footage, Polar Bear trails were visible to the reviewer. Yet previously, live on-screen images during the sampling stage showed no trace of the bears or their trails. If they had relied on their field findings alone, they would have reported no bears (a false negative in this case).

In their study of deer, Preston (2021) used a team of 'observers' to analyse still thermal images generated by a FLIR Vue Pro R drone-mounted camera. Their manual analysis procedure involved sifting through still images until they identified a potential target which was then classified as deer or non-deer. They attempted to use subcategories relating to surveyor confidence but found it was variable and subjective. They suggest that the wider applicability of the method is dependent on the research and development of improved image analysis procedures (by manual or automated methods).

Inter-analyst variability can present issues relating to data consistency. When Hambrecht et al. (2019) used a small team of analysts, they found a significant effect on detection probability. They state that this may be due to the different levels of analyst experience. While this study tested the use of drones equipped with both thermal and RGB cameras to detect poachers rather than wildlife, the principles of detection are similar, so this should be considered when using multiple analysts for wildlife projects.

It is important to reiterate that manual analysis can suffer the same issues as any human observer might, including tiredness and inattention (Chretien 2016). Where it is not possible to automate the process, these issues must be mitigated as much as possible by implementing procedures such as cross-checking, quality assurance and time management to avoid screen blindness and fatigue.

Automated analysis

With the ever-increasing volumes of data we collect, automation of data analysis is becoming increasingly necessary for many applications. Automation of the processing and analysis of this data has the potential to revolutionise the field in a number of ways, including increasing reliability and even standardising the technique (Chretien 2016) by reducing the variability that is naturally associated with manual analysis. However, the nature of thermal-imaging wildlife data poses unique challenges to automation.

One major challenge here is fulfilling the requirements of data input for automation programmes. Among other criteria, automation processes may require high-quality data, the stillness of the background in the frame and low signal-to-noise ratios. This should be considered carefully during the planning stage (see Section 3.1.3) so data collection at the sampling stage (see Section 3.1.5) can be carried out to meet these criteria. Where this is not possible, as was the case in McMahon et al.'s study (2021), it may be necessary to switch to a manual data analysis procedure. Using drones to detect and monitor Moose *Alces alces*, the authors reported that the thermal-imaging data they acquired was unsuitable for the automated process they planned to use. This issue was largely attributed to the detector resolution of the thermal device they used and their flight height limitations to minimise potential disturbance issues.

There are a number of different ways to detect and track animals automatically using algorithms. These include:

- deep learning
- image recognition
- adaptive thresholding
- background subtraction

Each of these methods operate in different ways and have been used for wildlife applications with varying degrees of success. Most automation procedures have been conducted using programming languages and environments such as MATLAB, Python and R. However, another approach is to use Geographic Information System (GIS) software. In order to automate the detection and counting of Grey Seals from thermal-imaging data,

Case study: ThruTracker (automated analysis software)

ThruTracker is an exciting new tool for the automation of video-based animal tracking (Corcoran, A. J. et al. 2021). Developed by Dr Aaron Corcoran and Dr Tyson Hedrick, this computer software programme promises to be an accessible yet powerful solution to the growing volumes of wildlife data requiring analysis. It can facilitate efficient detection, tracking and visualisation of wildlife movement from thermal video data.

How does it work?

The software itself is coded in MATLAB (MathWorks, Natick, MA, USA) but can be accessed freely in an app-based environment without the need for paid licences. It uses a background subtraction algorithm to detect and track moving objects which can then be visualised in either two-dimensional or three-dimensional space depending on the data input.

Can I access it?

Unlike other programmes that have been developed, ThruTracker is freely available (open-source) and does not require specialist coding skills or expensive software packages. Currently, you can access the software (source code download via GitHub), user manuals and video tutorials online. For those who have the skills to do so, the code itself can also be customised (conditions apply).

What can it be used for?

The software is designed to detect and track moving objects, so is most applicable for wildlife that is in motion at the time of sampling. With this condition in mind, bat and bird species applications are among the top candidates for optimum use (for details see Case Studies 1 and 2 presented in Corcoran, A. J. et al. 2021).

For more information on ThruTracker, you can visit the Sensory and Movement Ecology Lab at UC Colorado Springs' ThruTracker webpage at https://sonarjamming.com/thrutracker/

Seymour et al. (2017) created models in ArcGIS. GIS software packages are regularly employed in wildlife and ecology, so where possible, they may present a more accessible approach to automation.

Semi-automated analysis

Semi-automatic analysis procedures harness the power of automated detection algorithms to save valuable time, and later employ manual inspection to improve accuracy or to achieve classification/identification. This powerful combination can facilitate data outputs that would be tedious and unrealistically costly to achieve through fully manual analysis, but that currently cannot be achieved using automatic procedures alone.

A semi-automatic analysis procedure was used by Schirmacher (2020) in their application focusing on bats as their target species group. In order to assess the efficacy of Ultrasonic Acoustic Deterrents (UADs), they used multiple thermal-imaging cameras to record bats at wind turbines sites. They later used multiple techniques to process and analyse the data they collected in the sampling phase. As part of their analysis procedure, they initially used an algorithm in MATLAB to detect moving objects (including bats, birds and insects). Following this initial automated process, they later manually reviewed the objects detected and classified them based on key characteristics, including shape, wingbeats and flight patterns. Using this process, they aimed to manually filter out non-target groups (birds and insects) and retain only data featuring the target species group (bats) for further behavioural analysis. A semi-automatic procedure, when used appropriately, can save considerable amounts of analysis time. As described above, manual analysis of this kind of data can be extremely tedious and time-consuming. An analyst may spend hours watching 'nothing' until a potential target appears. Semi-automatic processes can leverage automated object detection to avoid this 'wasted' analyst waiting time. Instead of watching and waiting for a target to classify, the automatic part of the process enables the analyst to work more efficiently: completing only the manual checking, classification and interpretation of behavioural data.

3.1.7 Reporting stage

Communicating the results of our thermal-imaging work may take various forms depending on the project or study objectives. This may be a scientific paper to be published in a peer-reviewed journal, a professional report to communicate to others within an organisation or to external stakeholders or a white paper. Whatever the specific form that our reporting takes, it is important to create a written document that effectively communicates the work that we have done. Reporting should be carried out by someone who has the relevant knowledge and experience to achieve this.

At a basic level, a document reporting thermal imaging used for a wildlife application should usually include:

- **Rationale** Explaining why we have done this work.
- **Methods** Clearly describing how we have done the work.
- **Results** Showing our findings (data) simply without interpretation.
- **Conclusions** Explaining our interpretation of our results and its implications.
- **Summary** Rounding up all of the above using accessible language.

If we cannot effectively convey these to our reader(s), we have gone to a huge amount of effort for little to no gain. The target reader may be a peer, a client, a government official, a decision maker or a member of the public. It is important to keep in mind the most appropriate language and levels of explanation that might be required for the specific reader or audience for whom the report or document is intended. Unnecessary levels of complexity or technical jargon may lead to confusion, disinterest or disengagement from the reader. Therefore, it is pertinent to carefully consider and select the appropriate language in accordance with the intended reader. When writing for peers in similar disciplines or specialities, discipline-specific technical language may be suitable. However, for more general audiences or for specific readers who are not technical specialists in this area, a different approach is needed. In either case, a Glossary of Terms is often helpful to assist the reader.

When reporting on the methods used in thermal-imaging wildlife projects or studies, there are some key details that are essential to include in order to provide a useful account of how it has been carried out. These are often (frustratingly!) missed in scientific papers and applied ecology reports, which makes it difficult for the reader to fully understand the procedure.

The first set of critical information the reader needs to know are the key specifications of the thermal-imaging equipment used, such as:

- device type
- make
- model
- thermal detector resolution
- video refresh rate (if collecting video data)
- sensitivity
- lens(es) used and resulting field of view (FOV)

It is important to clearly describe how this equipment was used by including details of:

- deployment method (e.g. drone-mounted or ground-based stationary)
- type of data collected (e.g. still radiometric images or non-radiometric video)
- operator information (including skill level, experience, certifications and training completed)

If post-processing and/or analysis was performed, include details of the procedure:

- analysis type: manual, automatic or a combination of both
- details of software used (including the specific version)
- analyst information (including skill level, experience, certifications and any project-specific training received)

3.1.8 Review stage

After a project or study has been completed, it is important to review and reflect on the processes and outcomes of each of the previous stages. During this stage, we can extract important information that can help refine methods to improve future work. Where possible, discussing the findings of a review with other wildlife professionals can not only help us to continue to develop and improve methods collaboratively but can also help others gain a better understanding of how they might improve their own work.

3.2 Levels of use

Different applications will require methods of varying levels of use or complexity. What is required will depend on the specific application and the types of data output required. Some applications will be relatively simple whereas others might be much more complicated.

At the simplest level, we might use thermal imaging as an aid to improve the detection of a target species in the field, without any recording of thermal-imaging files (see D in Figure 11). Moving up a level, we can add in the recording of non-radiometric thermal images or video (see C in Figure 11) allowing us to go back and spot check to verify our observations in the field. At the next level, we also record non-radiometric data but this is analysed in full, not spot checked (see B in Figure 11). At the highest level, we record radiometric thermal images or video to be processed and analysed after sampling is complete (see A in Figure 11). As we move up the levels from D to A, not only do we increase the complexity of use, we can also increase the level of accuracy that is possible (as long as we meet appropriate conditions relating to species application, environmental conditions, etc.).

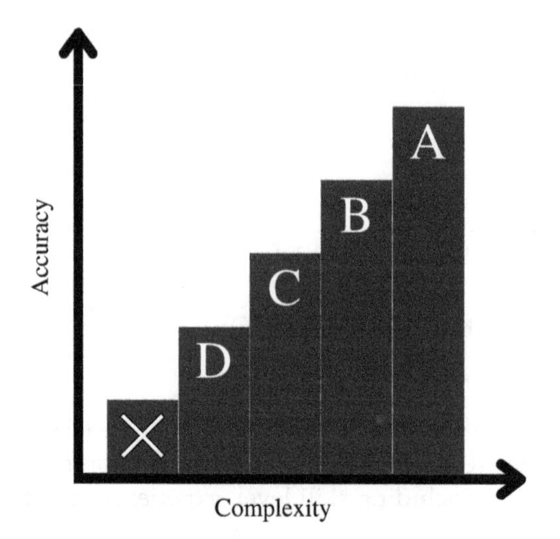

Figure 11 Represents the main levels of use: as we step up the levels from D through to A, potential accuracy generally increases with increasing technical complexity. The white 'X' represents thermal not being used (i.e. traditional visual methods) for the purpose of context.

3.3 Deployment methods

There are lots of different ways to deploy thermal-imaging equipment and in this section, we will explore some of the main options. Deployment of thermal-imaging devices or systems can be:

- ground-based – stationary
- ground-based – roaming
- aircraft-mounted
- drone-mounted
- on board watercraft/water vessels
- vehicle-mounted
- vantage-point-based
- deployed at height using machinery
- pole- or scaffold-mounted

Each of these deployment methods come with their own set of benefits and challenges that should be carefully considered during the planning stage (see Section 3.1.3).

3.3.1 Ground-based – stationary

Stationary ground-based systems or devices are usually attended by camera operators but could be deployed to record automatically (though this is not common practice at present due to the value of equipment and the skills required to set up such systems). Brawata et al. (2013) set up a custom-built system to automatically detect and record the behaviour of carnivores in the field. Their system included a FLIR ThermaCAM S45 connected to a laptop and a power supply connected to an inverter and 12-volt battery. This system only recorded thermal video when triggered via their automatic thermal video capture system. The process used background subtraction and predefined outline parameters to begin recording when the target animal entered the camera's field of view. Because the system was to be left unattended during sampling, they protected vulnerable sections of their system in a water-resistant casing and a waterproof container.

Thermal-imaging devices for the former are commonly tripod-mounted to reduce or eliminate movement of the frame capture, to both secure the device and to ensure image quality to allow for later review and/or analysis (see more on tripods in Chapter 4).

Ground-based tripod-mounted cameras can be oriented in any direction, according to the target species. In their study on the flight directions of migrating birds, Anderson et al. (2021) oriented their camera vertically to record the flight direction of birds flying directly overhead.

3.3.2 Ground-based – roaming

Some thermal-imaging devices are designed to be handheld, and if they are lightweight and ergonomically suitable, they can be used for roaming ground-based applications where the user may walk with the device in a defined area or along a transect. However, this option can be problematic for health and safety reasons. Using any technology with a screen requires care: it can have implications for optical health when used for long periods and can also distract the user from hazards in the environment. The ergonomics and weight of the device should also be taken into account. Many devices that are marketed as 'handheld' can actually weigh over 2 kg (alternative lenses can further add to this weight) and thus would not be safe to carry around for extended periods. Therefore, for this deployment option, it is important to properly evaluate devices on a case-by-case basis and select lighter devices that can be easily handled and operated in the field.

Medolago (2021) used a FLIR T650sc thermal-imaging camera during walked transects as part of their study of technology to detect birds at an airport. The camera operator walked on foot and scanned the area at regular (50 m) intervals. Though the authors of the paper do not discuss this, this camera model is both heavy and expensive in comparison with other available options. It should be noted that walking around with this device for long periods, especially in the dark, is risky in terms of potential damage to the device and to the operator. If you are thinking of following a similar procedure, it is advisable to consider a more appropriate device.

Bedson et al. (2021) used a thermal scope with binocular output while searching for Mountain Hare *Lepus timidus* using a point transect method. This involved night-time surveys in total darkness, where observers walked set transects and then set up the thermal-imaging device on a tripod to conduct 360° scans of the area. The Armasight Command 336 HD device they used was a ruggedised device with a waterproof enclosure designed for military and hunting applications, features which lend themselves to a roaming application in the field.

3.3.3 Aircraft-mounted

Aircraft deployment can be useful for the survey and monitoring of some species and has the advantage of allowing the coverage of large areas in a relatively short space of time. However, the use of aircraft for wildlife has its difficulties, including the high costs and safety concerns. Aircraft accidents have even been previously reported as one of the leading causes of death in wildlife professionals (Sasse 2003). When using thermal imaging for wildlife, aerial surveys are generally reported as useful for larger species in more open habitats, whereas complex habitats can be much more challenging. Altitude and equipment specifications can play a large role in determining the usefulness of this mode of deployment. To ensure imagery is useful for wildlife applications, it is necessary to use stabilisation of the device during sampling.

3.3.4 Drone-mounted

Drones, also known as Remotely Piloted Aircraft Systems (RPAS), Remotely Piloted Aircraft (RPA), Unmanned Aerial Vehicles (UAV) and Unmanned Aircraft Systems (UAS), have become popular in recent years for both applied and research-based wildlife applications (Hyun et al. 2020; Fust and Loos 2020). Some drones with integrated thermal-imaging devices are now commercially available, while others may have gimbals allowing the option to mount a range of separate devices.

In recent years, this approach has become very popular because of the range of benefits it can bring to wildlife data collection (Seier 2021). Drones can allow us to collect data in a quick, efficient and cost-effective manner which often outperforms conventional methods in these areas. In some situations, they can provide an alternative aerial approach to aircraft at a lower cost and risk to human life. Routes can be stored for long-term use, allowing for the monitoring of species in a repeatable manner that can allow for the comparison of data over time (Kim 2021).

However, drone deployment comes with its own unique set of challenges. From a practical perspective, they require a unique set of skills to operate the equipment safely and efficiently. In most countries, their use is restricted and is subject to airspace restrictions (Chabot 2015). It is important to ensure you follow local regulations and legal requirements according to the country you plan to operate in.

The use of drones may cause disturbance to wildlife, which may affect the target and non-target species during deployment. This should always be taken into account and mitigated accordingly (Seier 2021).

As with aircraft deployment, stabilisation should be used when using drones to acquire thermal-imaging data to maximise image quality and minimise issues associated with vibration or motion (Chretien 2015). Testing (see Section 3.1.4) can be crucial for novel drone-mounted applications. Having experienced both challenges and benefits in their study using thermal drones for moose detection, McMahon (2021) advised the use of test flights prior to sampling to minimise issues that might hinder the success of such applications.

3.3.5 On board watercraft

Studies of cetaceans have unsurprisingly involved the use of watercraft (Cuyler 1992; Baldacci et al. 2005; Burkhardt et al. 2012; Zitterbart 2013). They include the use of a 30 m boat (Cuyler 1992) but more often involve much larger research ships or vessels (Baldacci et al. 2005; Burkhardt et al. 2012; Zitterbart 2013). This approach is not just for marine use, however: it has also been used for the deployment of thermal imaging to detect primates in tropical rainforests (Jumail 2020).

3.3.6 Vehicle-mounted

Vehicles have been used in a number of studies using thermal imaging for wildlife applications. Devices have been used from inside and outside of the vehicle. Although a convenient way to cover larger areas, safety should be the first consideration when planning this deployment method.

Studies on the European Brown Hare *Lepus europaeus* have used vehicles to conduct driven transects using thermal imaging (Karp 2020; Sliwinski et al. 2021). One study incorporated assessment of thermal imaging and detection dogs as methods to detect hares in Switzerland (Karp 2020). Using a handheld FLIR Scout TS-32r Pro from a custom-made platform on a pickup truck, the viewer scanned the area vertically as their driver operated the vehicle on their sampling route. Sliwinski et al. (2021) conducted a comparative study on the use of thermal imaging vs traditional spotlighting techniques, using vehicles to cover large areas of agricultural land as they searched for their target species. During sampling, the observer stood on the loading area of a pickup truck (secured by a harness) while being driven at speeds of 5–10 km/hour. It was not clear whether the device used (Nyxus Bird) was handheld or secured to the vehicle, but further investigation of the device specifications suggests that either method would have been possible (Infratec 2021).

In their UK-based study on deer detection methods, Gill et al. (1997) found that vehicle census methods were as effective as those completed on foot but took considerably less time. Their vehicle-based sampling was conducted from an open-topped vehicle via transect routes that were almost entirely closed to the public. They used a specialist thermal imager designed for weapons targeting (Pilkington Thorn Optronics, LITE), which allowed them to detect their target species up to 2 km away.

Devices can also be used from the interior of a vehicle. However, because glass reflects infrared waves, this must be done through an open window. Alternatively, one could use a thermal window (see Chapter 4 for more details on IR windows). Zini et al. (2021) used a handheld thermal-imaging scope (Pulsar Helion XP50) from the open passenger window of their vehicle to detect deer in Thetford Forest in the UK. They used driven transects to acquire data on deer numbers and used brief spotlighting with a lamp and laser range-finding binoculars to sample the distance between the researchers and the target animal. It should be noted that this latter use of 'lamping' can cause disturbance to wildlife. This is largely unnecessary and can be avoided by using alternative equipment (stand-alone range-finding devices or thermal-imaging equipment with inbuilt laser rangefinders are available).

Taking a somewhat similar approach to Zini et al. (2021), McGregor et al. (2021) used a FLIR III 640 thermal-imaging scope operated from the rear passenger seat of a vehicle. However, in the latter study, another researcher operated a spotlight from the front passenger window. With both devices facing in the same direction, they attempted a side-by-side comparison of thermal-imaging and spotlighting methods.

Occasionally, wildlife seekers manage to acquire specialist vehicles equipped with their own built-in thermal-imaging cameras. Usually designed for military use, these are rugged machines capable of accessing off-road terrain. In their programme on *The Great British Year*, the BBC's film crews obtained this type of vehicle to film wildlife. Modern civilian vehicles are also being equipped with a range of different sensors to improve human safety and move towards autonomy (Silva 2021). Perhaps one day we might be able to use thermal imaging as an integrated field vehicle system.

3.3.7 Vantage point-based

Vantage points can provide an excellent opportunity to set up thermal-imaging devices to gain clear views of areas of interest. Where available, this can negate the need for aerial or elevation equipment. It is worthwhile to search for possible vantage points at the planning stage and ground-truthing their usefulness during the testing or pilot phase.

3.3.8 Deployed at height using machinery

There are a range of different machine-based platforms that can help deploy thermal-imaging technology in situations where elevation is needed. Mobile elevating work platforms (MEWPs) allow us to deploy both thermal-imaging equipment and operators to defined heights as needed. When selecting a MEWP, it is important to ensure that the selected machinery is suitable for the height you wish to reach, allows an adequate and safe working area for your set-up (and any operators if attended) and minimises the movement in the frame of the thermal-imaging device. MEWPs must be operated by a suitably qualified professional, and this may require safety certification for individuals controlling the MEWP and those working on the platform itself (for more details check with the relevant regulations in the country you work in).

3.3.9 Pole- or scaffold-mounted

You could also mount your thermal-imaging device onto a pole or scaffold, though there do not appear to be any examples of this in the field so far. If you do decide to try this, consider the stability of the structure and how this might affect your imagery. Using particularly long cables can affect data acquisition, so make sure you take this into account if using them.

3.4 Combining techniques

For some applications, using thermal imaging in combination with other techniques can maximise our output. This might include a combination of other imaging technologies such as RGB or multi/hyperspectral cameras. It may also include different modalities to complement imaging data, such as collecting sound data using acoustic recording systems alongside thermal-imaging systems.

Approaching data collection with a combination of techniques can help us to piece together the puzzle, allowing us to gain a richer understanding of the wildlife scene. It can help us 'fill in the gaps' where a single technology cannot fully answer our questions. Rahman and Rahman (2021) evaluated the use of drone-mounted thermal-imaging devices, camera traps and traditional transect surveys for the detection of a range of wildlife species in Indonesia. They concluded that while each tool has benefits and drawbacks, using two or more methods in combination can improve the overall outcome.

While combining techniques can be advantageous, this approach requires careful planning to be deployed effectively. It is important to consider:

- how different systems might affect/interact with one another;
- if time synchronisation is required and if so, what temporal scale is acceptable/appropriate;
- how the different modalities compare in terms of detectability.

One common combination of techniques is the simultaneous use of thermal-imaging and RGB or visible-light cameras. Most often, this utilises the thermal component for detection of the target species and relies on the RGB/visual data for further identification or other information not available or not easily seen from the thermal data. Florko et al. (2021) used a FLIR T1020sc thermal-imaging camera and a Nikon DSLR camera on board a light aircraft to successfully detect and study Narwhal *Monodon monoceros*. This combination approach allowed the effective detection of 'flukeprints', as these marine mammals breach the surface to breathe, while the use of the DSLR imagery provided important details for species identification and counting the number of animals present. When Hambrecht et al. (2019) tested drones with thermal and RGB cameras to detect poachers, they found that thermal imaging had a higher detection probability than their RGB cameras overall. However, under daylight conditions with higher temperatures, the RGB camera outperformed the thermal camera. Hence, employing two different sorts of camera may allow us to switch between using one or the other, depending on the conditions.

Thermal imaging has also been used in combination with radiotelemetry to precisely locate wildlife. In order to locate Capercaillie, Fletcher and Baines (2020) employed radiotelemetry to find the rough location of female birds fitted with radio transmitters and then implemented a thermal-imaging scope to hone in on the exact location of the target animal (for more details on this study see Section 6.2). When used for bats, thermal imaging is usually employed in conjunction with acoustic devices to simultaneously record their ultrasonic echolocation calls (Fawcett Williams 2021). This allows us to gather information about the number of bats, their behaviours and flight patterns from the thermal-imaging data and then extract information on the likely species present from the acoustic data.

Dual-use is not the limit of course: we can bring in a multitude of different techniques and technologies to pull together different pieces of our wildlife puzzle. Diehl et al. (2015) evaluated a range of techniques to detect bats, birds and insects at solar energy towers. Their evaluation included the use of five different camera systems using different wavelengths, with thermal imaging among them.

3.5 Expertise requirements

Thermal imaging is a technical discipline and requires a combination of specialist knowledge and skills to be applied successfully for wildlife applications. Each wildlife application requires an appropriate level of expertise in order for operations to be carried out effectively.

From a thermography perspective, these expertise requirements can be divided into three main categories depending on whether they include:

- detection
- temperature measurement
- health assessment or diagnosis

For wildlife applications that rely on thermal imaging for the detection of wildlife, it is important to ensure that this work is carried out effectively. To facilitate this, those responsible for the application of thermal-imaging technology must be equipped with the necessary knowledge and skills to do so. Havens and Sharp (2016) emphasise that users must be familiar with the operation of thermal-imaging devices in order to obtain quality imagery. They advise that training is necessary prior to attempting fieldwork, so that individuals can become familiar with the use of equipment and gain an understanding of how to optimise their methods. Goodenough et al. (2018) state that for real-time thermal imaging to be used as a reliable technique it is essential that users are provided with tailored training. However, it is important to add to this excellent advice that expertise is not only required for field deployment at the sampling stage, but it is also necessary to effectively plan, analyse and report as well. At the most basic level, those involved must understand how the technology works, its inherent benefits and limitations and how to use it correctly for the specific application. Such expertise can be developed through wildlife-specific thermal-imaging training courses or programmes, which will vary depending on the wildlife species involved (see the Supplier Directory on page 127). For bat survey applications, you can find specific guidance on expertise requirements within the Thermal Imaging: Bat Survey Guidelines (Fawcett Williams 2021).

Thermographic training is necessary for research and other specialist applications where temperature measurement is required. A Thermography Level 1 Certification training course is a good place to start. Alternatively, some universities may provide training within their own departments.

Wildlife health assessments should always be overseen by an appropriately qualified and experienced veterinary professional. Diagnosis must only be carried out by a licensed veterinarian. However, camera operators or technicians with the relevant expertise can collect thermal-imaging data to inform assessment or diagnosis by a veterinary professional who is experienced in using and interpreting thermal-imaging data for these purposes. Extensive specialist training is required to gain the knowledge and skills required to produce thermal images that are of a clinical standard.

You can find details of training providers specialising in detection, temperature measurement and health assessment applications in the Supplier Directory.

3.5.1 Additional expertise requirements

Once thermal-imaging-specific expertise requirements are taken into account, there may also be other expertise and/or certification requirements for some applications. Often these are for the more specialist projects including drone or MEWP operations where country-specific regulations are often in place. It is important to check what the local requirements are so that you can carry out your work in line with legal and/or professional guidance as necessary.

Project-specific training can also be extremely valuable, especially where large teams are involved. The value of this is particularly evident from the successful use of this technology in the long-term volunteer-based study of Hedgehogs in Regents Park (Bowen et al. 2020; Gurnell et al. 2021). With the help of experts in thermal imaging, the authors of this study designed a standardised protocol to incorporate the use of this technology as part of a wider and similarly well-designed project. For some applications, especially novel ones, bringing in specialist expertise can also be hugely valuable.

Kayleigh says:

If you are thinking of attempting to use thermal imaging without acquiring the right expertise, it is important to think about the consequences of getting it wrong.

I don't think many untrained users realise what they could be missing when attempting to detect wildlife with the equipment they have. When it does go wrong, lots more can go wrong later down the line. However, no matter how much I reiterate this point, I don't think anything really beats seeing it.

Take a look at Figure 12 for a classic example. If the 'no bats' data shown was reported in this situation, the potential consequences include fines, spiralling project costs, major project delays and the disturbance, injury, or death of bats.

This example is taken from the Analysis Stage (see Section 3.1.6) but errors can occur at any point in the workflow process.

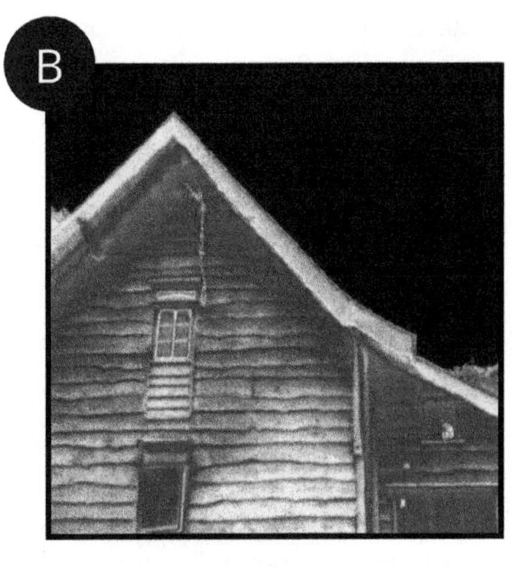

Figure 12 Processed thermal images derived from the same original radiometric thermal image which has been processed manually (A) with and (B) without appropriate expertise. (A) clearly shows the target animal (bat) against the background, whereas in (B) the bat has 'disappeared' due to inappropriate use of software settings.

4. Equipment

For any wildlife technology application, it is essential that we choose the right tools for the task at hand. This is particularly important when using thermal-imaging technology (Figure 13 illustrates just how different equipment set-ups can be). In this chapter, we will look into the different types of thermal-imaging device, and their key technical specifications. We will also highlight some other related considerations that may be necessary to implement this technology for your specific wildlife application.

4.1 Important specifications for device selection

To select appropriate thermal-imaging equipment for a wildlife task we must examine the specifications of potential devices. We will go through some key specifications that we can use to determine a device's suitability in the following sections. Each of these should be detailed in the technical specifications for the device (sometimes referred to as a datasheet) or equivalent. This can usually be found on the manufacturer or supplier's website as a downloadable PDF document. When assessing a device it is important to refer specifically to these technical specifications and not to rely on details provided in marketing copy, as these can be confusing or even sometimes misleading to those who are not thermographers.

Figure 13 Different wildlife applications require different thermal-imaging equipment: (A) shows a handheld thermal-imaging scope used for real-time detection in the field, (B) shows a thermal-imaging camera system with a laptop to enable recording for later analysis.

If you are unable to select a suitable device with confidence, you may need some assistance. In this case, it is advisable to consult a certified thermographer with relevant wildlife expertise who can offer professional advice to help you (see the Supplier Directory on page 127).

For detailed information on equipment specifications for bat survey applications see Fawcett Williams (2021).

4.1.1 Device type

There is a broadening array of thermal-imaging device types that are becoming available to us at increasingly affordable prices. However, few of these devices have been designed specifically for wildlife applications and not all devices on the market are appropriate for use in this field. Likewise, not all thermal-imaging devices are suitable for wildlife detection specifically (Boczkowski 2021) so it is important to ensure that we evaluate potential devices with this in mind. In order to be able to collect high-quality data, it is vital that we choose suitable equipment in line with the specific wildlife task at hand.

Devices range in size and complexity: from small, low-resolution smartphone attachments to high-definition SLR-style camera models (such as that shown in Figure 14) and beyond. Prices start at three figures for a small handheld unit, to five figures for higher end models (based on UK commercial prices in GBP, early 2022). That being said, prices have decreased dramatically in recent years and it is likely that economies of scale will continue to make equivalent devices more affordable as time goes on.

Often more complex devices offer us a wider range of features that can allow for higher levels of accuracy. For example, a simple device may allow direct viewing only with no recording capacity, another may allow non-radiometric recording to allow us to check footage but not optimise it thermally. A higher end device with radiometric recording can

Figure 14 A FLIR T1030sc thermal-imaging camera set up for a bat survey of a building. This is an SLR-style device with an integral display and switchable lenses.

allow us to optimise and analyse data to extract data with much higher accuracy; however, this improvement in accuracy comes with a cost of system complexity, time, expertise and financial cost of the device. For more on radiometric versus non-radiometric data outputs, see Section 4.1.7 below.

4.1.2 Detection distance (range)

If it is available, the **detection distance** (or the **range**) of a thermal-imaging device is usually detailed in metres (m) from the device. The range is the parameter usually reported for commercially available devices and these are usually based on the size of a human, or sometimes a deer under specific thermal and environmental conditions (which are not always specified).

Detection distance is defined principally by the:

- size of the object or animal
- detector resolution
- lens used
- effects of atmospheric conditions

This means that for species-specific detection distances, we can calculate the expected detection distance using these key device specifications, along with likely thermal and environmental conditions and the body size of our target species. These calculations are usually relatively conservative, but they should always be ground-truthed during testing (see Section 3.1.4) as the detectability of animals in the field can vary due to a range of factors.

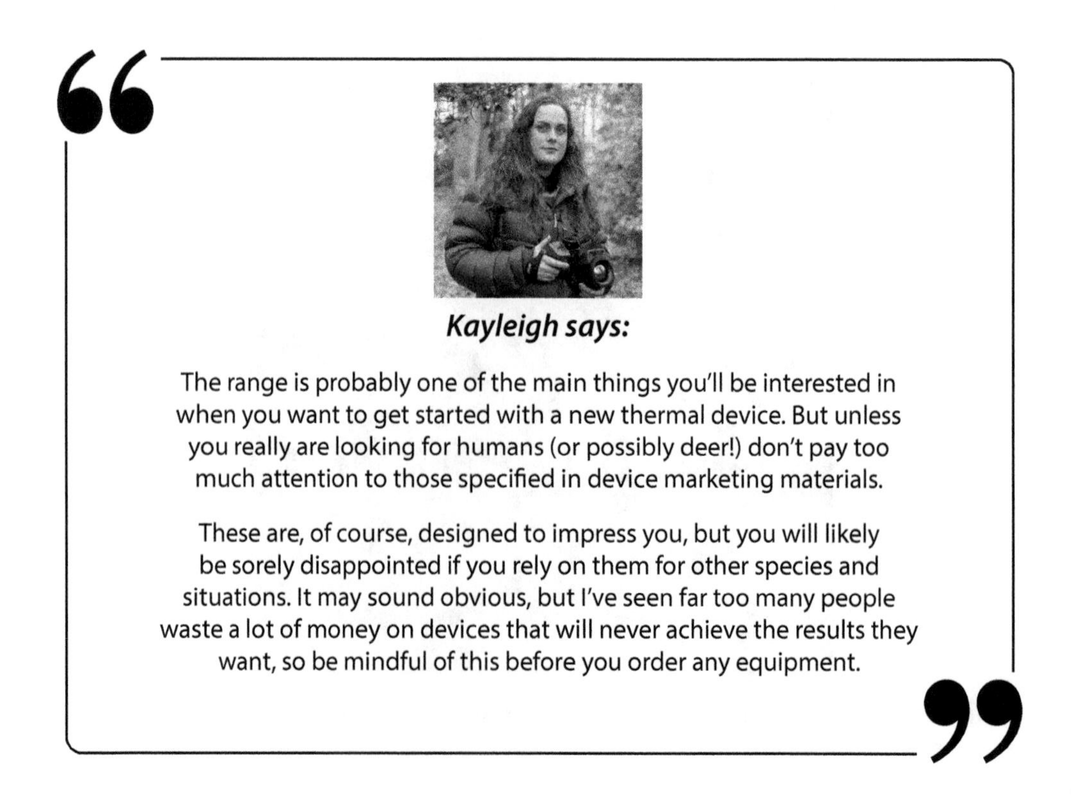

Kayleigh says:

The range is probably one of the main things you'll be interested in when you want to get started with a new thermal device. But unless you really are looking for humans (or possibly deer!) don't pay too much attention to those specified in device marketing materials.

These are, of course, designed to impress you, but you will likely be sorely disappointed if you rely on them for other species and situations. It may sound obvious, but I've seen far too many people waste a lot of money on devices that will never achieve the results they want, so be mindful of this before you order any equipment.

4.1.3 Spectral range

The thermal detectors inside thermal-imaging devices are selectively sensitive to specific bands of infrared waves (shown in Figure 15). This defines the device's spectral range band, and thermal-imaging devices suitable for wildlife applications fall into two spectral band categories:

- Long Wave (LWIR) 8–14 μm
- Medium or Mid-Wave (MWIR) 2–5 μm

All of these operate in bands that use longer wavelengths than that of visible light (which spans 0.4–0.7 μm). In general, devices used for wildlife survey applications tend to fall within the Long Wave (LWIR) spectral range category which are sensitive to wavelengths between 8 and 14 μm. One advantage of these LWIR devices is that they are less sensitive to solar reflections (Öhman 2014), which can help reduce issues related to false positives from non-target objects in the field. Medium or Mid-Wave (MWIR) devices are sensitive to wavelengths between 2 and 5 μm. These have also been used for wildlife, but these types of device tend to be very high-end cameras more suited to research and development applications. However, provided that the other device specifications are adequate for your application, you can use either LWIR or MWIR for wildlife.

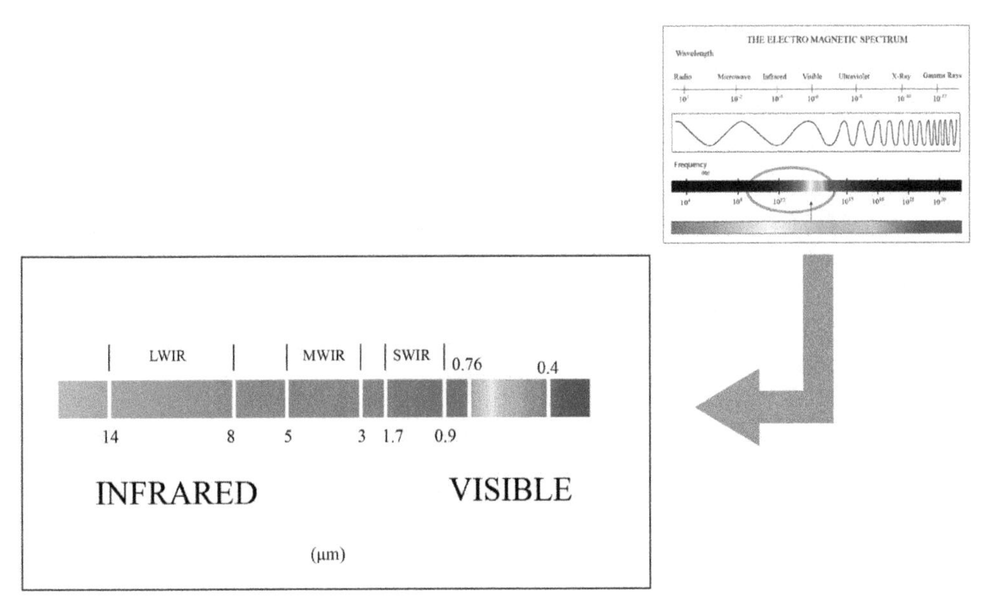

Figure 15 The locations of Long Wave (LWIR), Medium or Mid-Wave (MWIR) and Short Wave (SWIR) within the electromagnetic spectrum.

4.1.4 Temperature range

It is important to ensure that a thermal-imaging device can effectively detect and display the likely temperatures of the target species and their surroundings in order for us to image them. This should not be an issue, however, as almost all thermal-imaging devices have an optimal temperature range encompassing this. Usually devices with a range of –40 °C to +120 °C are used for wildlife applications. Wider temperature ranges are available and can be used for wildlife but the broadening of this capability is usually associated with higher device costs.

4.1.5 Detector resolution

The resolution of thermal-imaging devices available for civilian use is currently considerably lower than that of most conventional (visible wavelength) cameras on the market. This specification is usually reported as the number of **horizontal pixels × number of vertical pixels** (e.g. 640 × 480 or 1020 × 768 are common detector resolutions of devices used for wildlife applications currently).

In combination with the lens used, the resolution of the thermal detector defines the range over which an animal can be detected. A detector with a higher number of pixels (higher resolution) will allow the detection of smaller objects or at greater distances compared to those with smaller numbers of pixels (lower resolution).

The achievable image quality is also limited by the detector resolution. This can have a crucial impact on the usefulness of a device for particular wildlife applications. For example, applications that involve the classification of animals to a group level or for species-specific identification require good quality images to provide the necessary level of detail. For applications where high image quality is required, we generally select the highest possible resolution available to us.

As with many features of technologies, however, there are some trade-offs to consider. First, increasing detector resolution generally comes at a proportionally higher purchase price. Therefore it is important to understand what the appropriate level of resolution for your application is to avoid unnecessary expenditure. Second, the size of the data output from thermal-imaging devices also increases with the image resolution. So to collect higher resolution data, you need to plan to handle and store larger volumes of data.

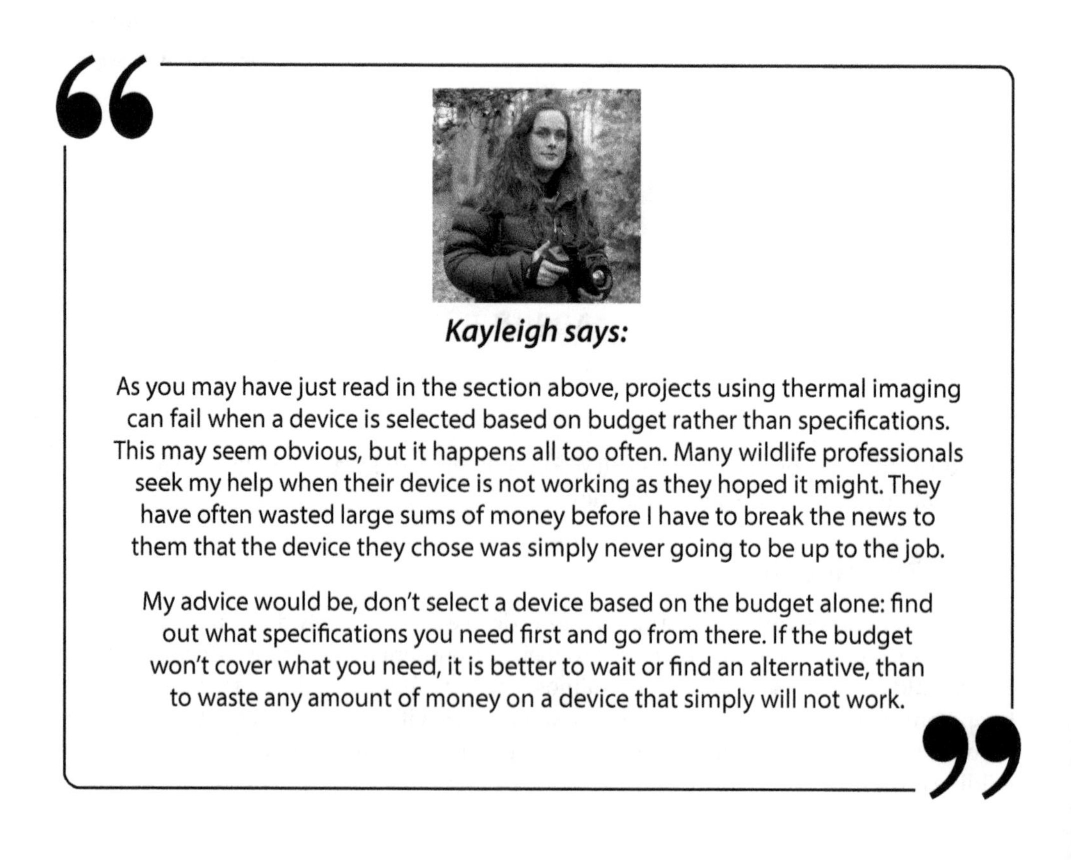

Kayleigh says:

As you may have just read in the section above, projects using thermal imaging can fail when a device is selected based on budget rather than specifications. This may seem obvious, but it happens all too often. Many wildlife professionals seek my help when their device is not working as they hoped it might. They have often wasted large sums of money before I have to break the news to them that the device they chose was simply never going to be up to the job.

My advice would be, don't select a device based on the budget alone: find out what specifications you need first and go from there. If the budget won't cover what you need, it is better to wait or find an alternative, than to waste any amount of money on a device that simply will not work.

This is particularly important if you are collecting video data, even more so if that data is radiometric, because the resulting files are especially data-heavy.

Selecting a thermal-imaging device with a detector resolution that is too low for your specific application leads to disappointing or even unusable results. For example, when Berthinussen and Altringham (2015) trialled thermal imaging as a method of monitoring bats around infrastructure projects, they used a device with a detector resolution unsuitable for their application. Due to budget constraints, they selected an unspecified FLIR device with a detector resolution of 320 × 240 pixels, which yielded very poor results. For an application like theirs, they needed a device with a minimum of 640 × 480 pixel detector resolution to detect bats over long distances.

4.1.6 Display

Digital displays are integral to most handheld or SLR-style thermal-imaging devices. Devices with displays sometimes allow this component to be turned on and off or to moderate the screen brightness. This flexibility can be particularly useful for wildlife applications where target species could potentially be disturbed by – or even be attracted to – the light emitted from the display (which can sometimes be very bright). It can also be useful for sensitive situations, such as when operating in areas where attracting unwanted attention from other humans could be problematic for both practical and safety reasons (such as monitoring wildlife at night in urban areas). Other types of thermal-imaging device may not have an on-board display component. These require an external display device to be connected, such as a laptop, tablet or a custom system that includes a monitor. Integral or not, the resolution of the display should have a resolution that is equal to or greater than that of the thermal detector if the full benefit of the detector resolution is to be realised there. For devices where this is not the case, the viewer will not see a true representation of the image that would be recorded digitally. This can have a number of implications depending on the way the device is to be used. For applications where the device is being used as an aid to visual detection in real-time in the field, it simply means that the resolution is not being used optimally. This may impact achievable detection distances or the minimum sizes of target species that can be resolved by the viewer. Display resolution can impact image quality due to focusing errors. When manually focusing thermal cameras in the field it may appear that the camera is in focus on a lower resolution display, when in fact it is not. Where possible, we can avoid this issue by using an external display with appropriate resolution to ensure that the focus is set correctly.

Some devices display images in false colour, some are limited to greyscale only and others allow the user a range of different colour palette options to choose from. Selecting an optimal colour palette can allow us to maximise the contrast that we can see in an image (Figure 16 shows how the colour palette can alter the appearance of a thermal image). Because human vision varies from person to person, this feature is very much a personal one and so it is important to determine the palette(s) that work best for the person who will be operating the device. If the device is to be used by more than one person in an organisation, not just as a device for a specific individual, it is important to consider the accessibility of the device for current and potential future users.

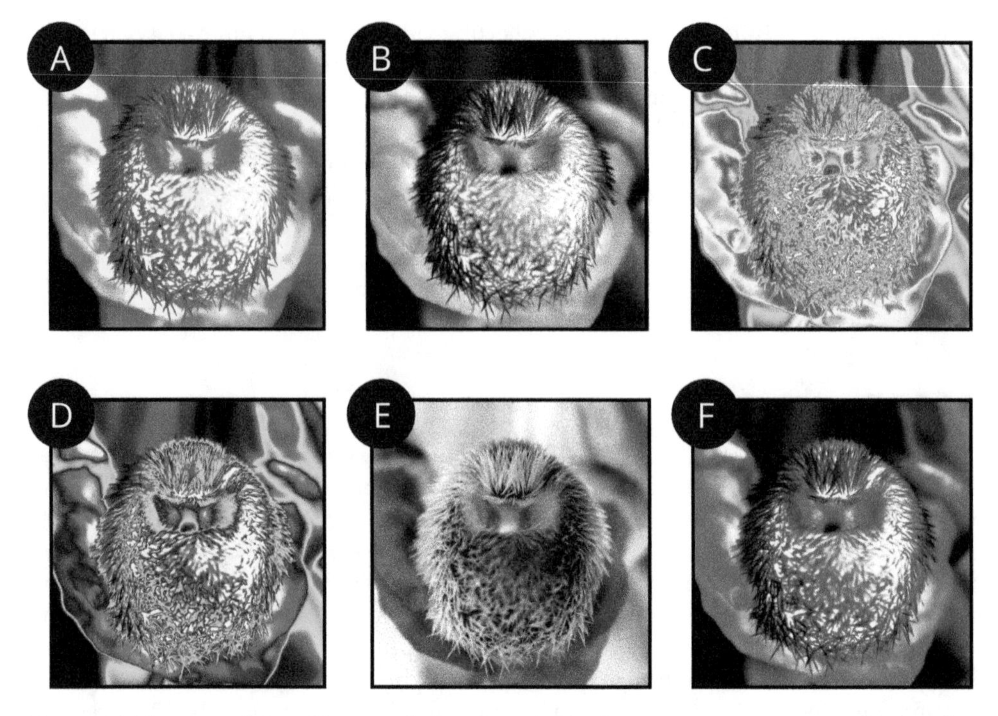

Figure 16 The same thermal image of a European Hedgehog *Erinaceus europaeus* displayed in a range of different colour palettes: (A) Ironbow, (B) White Hot, (C) Rainbow 2, (D) 1234, (E) Black Hot, (F) Blue.

4.1.7 Data outputs

At the simplest level, the data output from a thermal-imaging device might only provide a still image or a continuous stream of images for live on-screen viewing. Such a device might be used as an aid to an observer who translates what they perceive visually via the display into another format (e.g. written or audio notes, populating a proforma, etc.).

Other devices allow us to save or record still images and/or video files. While not all thermal-imaging devices have video recording capability, it can be essential for some applications. There are a number of benefits to having a video data record:

- They can be reanalysed to gain more information (e.g. videos initially used for count data could be reanalysed later for behavioural data).
- They can be automatically or semi-automatically processed. This can save a lot of time and money and allows for scalability.
- They can be cross-checked by multiple analysts. This is particularly important where images or video are difficult to interpret (Diehl et al. 2015) or for situations that require a particular level of expertise to classify correctly.
- They can be checked for quality assurance purposes.
- They can be checked in the event of an inquiry or investigation, and even be used as legal evidence.

Video refresh rates can vary, and it is important to make sure that this is adequate for your application. This parameter is the number of frames captured per second and is usually given in frames per second (fps) or in hertz (Hz). For bats and birds in flight, a minimum frame rate of 30 Hz is required for detection, although higher frame rates can provide higher quality footage (at a cost of increased file size).

Radiometric vs non-radiometric thermal-imaging data

Thermal image or video outputs come in either radiometric or non-radiometric formats. Radiometric outputs contain thermal data for each pixel within each image or frame of video. This means that the image can be processed, optimised and analysed in specialist thermal software to give high levels of accuracy. Non-radiometric data outputs, on the other hand, come in standard image or video file formats and are relatively light in terms of file size. These files contain only the colour or shade representations used at the time of recording and cannot be thermally processed later.

Non-thermal data outputs

As well as providing a thermal data output, some mid to high-end cameras may also have a number of other on-board sensors or components that allow them to provide additional types of data output. In some models, the integration of Global Positioning Systems (GPS) allows the generation of location data to coincide with the collection of thermal data. This could be useful for applications where sampling is conducted at multiple locations, as it allows the simultaneous collection of these two parameters without the need for a stand-alone GPS unit or similar. Likewise, as the data are already 'paired' it also streamlines the process of collecting and utilising the data by negating the need for matching up the two data types manually.

Some thermal-imaging devices also have built-in digital (visible wavelength) cameras. These give the additional output of photographs alongside paired thermal images taken at the same time. Some devices can also produce varying forms of hybrid images of thermography and photography. In situations where ambient light allows, having corresponding photographs can provide useful contextual information for later interpretation of thermal images.

4.1.8 Ingress protection

Most thermal-imaging devices were not designed for wildlife applications, so unfortunately they are not all as weatherproof, or perhaps 'field-proof', as we might like them to be. To help us gain a better understanding of how appropriate a given device might be for the environmental conditions we plan to use it in, we can refer to its International Protection (IP) Rating (International Electrotechnical Commission 2013). Defined by International Standards, these IP Ratings – sometimes also referred to as codes – classify the level of protection an electrical enclosure provides against external agents, including dust and water. You can find out more about how to read IP Ratings at https://www.enclosurecompany.com/ip-ratings-explained.php

If you find a device that fits your needs in terms of other critical specifications but it does not have the level of protection you need, all is not lost. You can consider the option of using additional housing for the device to provide the necessary additional protection to keep it working safely in the field. Housings can be either readymade or custom made for your device. Readymade housings are only available for a select number of devices. A custom-made housing is a good option provided that it is designed and produced to safely house the device without compromising its operation, for example not causing it to overheat. Whether readymade or custom-made, housings must be fitted with an IR window (standard glass must not be used) that is appropriate to the wavelength sensitivity of the device. If you are unsure about the suitability of weatherproof housings and/or IR windows for your chosen device, you should consult a certified professional thermographer for clarification.

4.1.9 Sensitivity

It is important to ensure that the device we choose is sensitive enough to allow us to discriminate between our target animal and the background habitat. Thermal sensitivity values, usually called Noise Equivalent Temperature Difference (NETD), tell us how small a difference in temperature the device can allow us to distinguish. NETD values are shown in millikelvin (mK) and the lower this value is, the more sensitive the device is.

4.1.10 Lens options: fixed or switchable

Lenses define the field of view within which we can sample and the detection distances we can achieve, so it is therefore critical to select the appropriate lens for your specific application. Some devices have one fixed lens that cannot be changed, while others offer multiple switchable lens options. Fixed-lens devices are usually lower in terms of purchase price, but offer no flexibility in terms of field of view, consequently limiting the different scenarios they can be used for. Devices with switchable lenses tend to be higher end devices with associated purchase prices, but can be better value because they can be used for a wider range of applications.

4.1.11 Uncooled or cooled

Cooled thermal cameras contain thermal detectors fitted with cooling systems that allow them to operate at cryogenic temperatures (FLIR 2021). This can allow them to produce high-quality images and video at high frame rates. Cooling mechanisms for these cameras can be noisy, which may cause disturbance to wildlife. These are generally only necessary for highly specialist applications requiring high temporal precision, such as determining wingbeat frequency in moths or behaviours that are extremely short in duration (e.g. a bat catching an insect). Since cooled cameras are unnecessary for the vast majority of wildlife applications, uncooled cameras are usually the more appropriate option.

4.2 Accessories

Some thermal-imaging devices function as stand-alone units, whereas others may require other equipment and/or accessories to function.

You may need to consider the following accessories:

- tripod(s)
- laptop(s)
- connections and cables
- power solutions
- weatherproof housing

- data storage
- travel solutions

Depending on your chosen device(s) it is important to make sure that you establish which, if any, of these are required, and to ensure they are budgeted for in advance and procured in time to begin testing and sampling.

4.3 Calibration

To ensure that a device is providing accurate data it should be sent for regular calibration checks as per the manufacturers recommendations (usually annually). A calibration check can be performed by the device manufacturer or by an approved independent laboratory, provided that the calibration details for the device are available.

5. Application Types

The number of wildlife applications used for thermal imaging is growing. In this chapter, we will explore the different types of application it can be used for.

Applications of thermal imaging for wildlife fall broadly into the following overarching categories:

- detection, classification and counting
- human–wildlife interactions
- behavioural studies
- health and rehabilitation
- thermoregulation and thermal biology
- communication, education and awareness

For each of these application types, we will explore some key examples where thermal imaging has been applied, the current understanding of its use and the implications or potential for use in future. For all of these applications, it should be noted that the results obtained by the authors in the studies and projects detailed in the following sections of this chapter were subject to the environmental and thermal conditions at that specific time and location. Transferring them for use in other situations should be approached accordingly and tested appropriately under relevant conditions before wider scale use (see Section 3.1.4 for more on testing).

5.1 Detection, classification and counting

Almost all wildlife-related projects generally require the detection of wildlife in some way. Once an animal is detected, it is also not unusual to then perform classification and, where possible, identification (Havens and Sharp 2016). In some situations, these tasks may be the sole aim of the work in question. For others, detection (and/or classification or identification) may be only one step or facet in a wider study or project. A great deal of effort can go into counting or estimating the numbers of animals in some way. Some projects or studies aim to make complete counts of individuals present (census). Others sample a subset of a population to infer or estimate the likely number of individuals in the population. Some simply aim to establish whether a particular species is likely to be present or absent. Long-term species-monitoring projects or studies rely on the successful detection of target species. Depending on the level of detail required and other factors, targets detected may also need to then be classified into groups or categories or identify them at species level. Methods for such projects should be repeatable so that they provide comparable data year on year (or at defined intervals over the duration of the project or study).

5.1.1 Signs

The majority of wildlife applications of thermal imaging focus on the detection of the bodies of individual animals of the target species. However, it is not always the animal itself that is the target for detection. There are some applications involving the use of thermal imaging that instead seek other signs of the target species that can indicate the animal's presence. These signs can come in many forms, including: tracks or trails (e.g. Brooks 1972), underground dens/burrows/warrens (e.g. Cox et al. 2021), nests (e.g. Benshemesh and Emison 1996) and other disturbances, patterns or evidence left in their surroundings (e.g. Cuyler 1992; Perryman et al. 1999; Florko 2021). A great example of these indirect indicators is the blow and flukeprints of cetaceans, which have successfully been detected for multiple species (Cuyler 1992; Perryman et al. 1999; Florko 2021). This is especially interesting and useful given the challenge water can present to thermal imaging for this group of marine mammals in particular (read more on this later in Section 6.4.4).

When using thermal imaging to look for signs of wildlife, it is important to remember that this technology does not allow us to see through solid objects or water. Variations or patterns in the apparent temperatures displayed in a thermal image can be caused by many different things. So-called hotspots or irregularities that can be flagged as signs of wildlife. Not all of these thermal anomalies are thermal signatures of animals. They may in fact be the result of other phenomena such as variations in material properties, including emissivity, surface geometry and reflectivity. So, it is important not to read too much into these features without a critical examination of what could be causing the variation in appearance; otherwise, we may well see patterns that are simply not there.

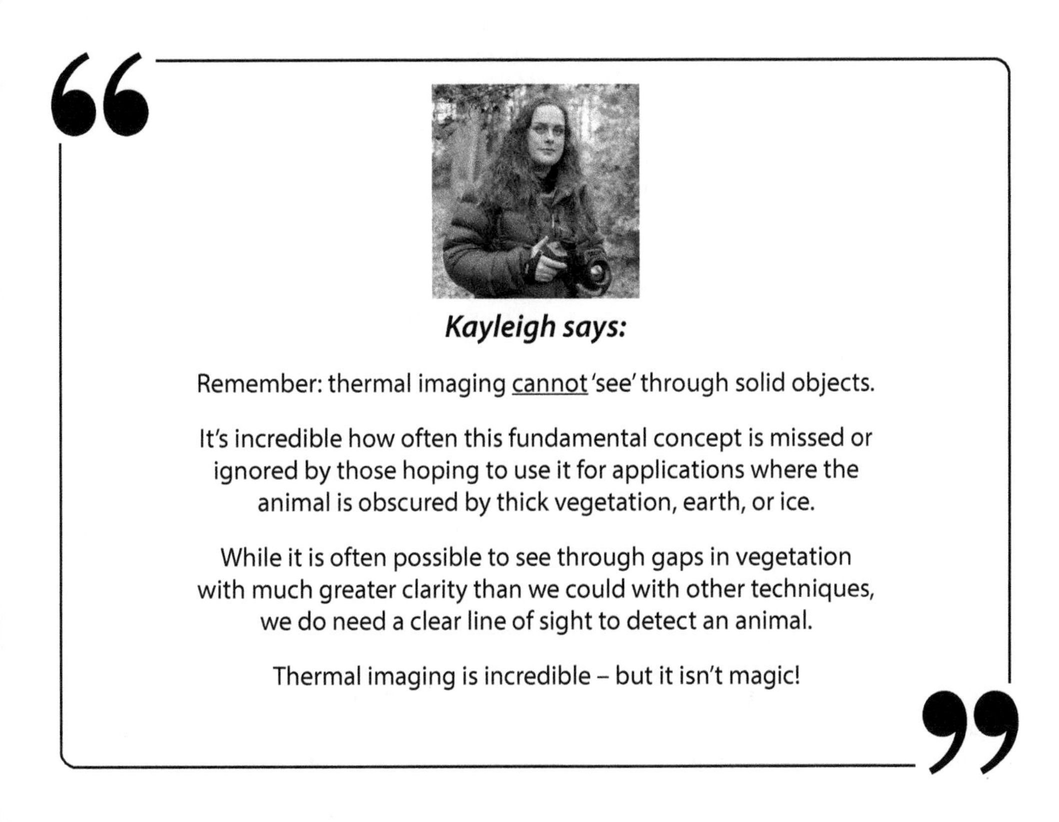

5.2 Human–wildlife interactions

As the growing human population encroaches further and further into the habitats of wildlife, the interactions between human and non-human species are increasing (Anand and Radhakrishna 2017). Many of these situations, often referred to as human–wildlife conflicts or interactions can be problematic for one or more of the species involved. We need a range of tools to help us avoid human–wildlife conflicts, mitigate issues arising from human–wildlife interactions and move towards a state of human–wildlife coexistence (Nyhus 2016).

Overexploitation or overharvesting by humans is the biggest threat to wildlife, followed by agriculture and urban development (Ritchie and Roser 2021; Maxwell et al. 2016).

5.2.1 Overexploitation

Direct exploitation or harvesting of wildlife species can be legal (hunting) or illegal (poaching), but both have potentially devastating consequences when wildlife species are overexploited.

Hunting

Humans hunt wildlife for a variety of reasons including: food, products (such as fur, bone, horn, etc.), recreation, pest control and population management for resource protection or conservation.

Human activities have resulted in the loss of countless wildlife species worldwide, including top predators that were once integral to most ecosystems. As a result, populations of deer and other large herbivores have swelled in the absence of their natural predators. Burgeoning numbers of these wild animals can have serious consequences for our habitats, economy and safety. In light of this, in many areas around the world it is now necessary to manage their numbers. In order to do this effectively, without tipping the balance too far one way or another, wildlife managers need to be able to detect and count wild animals to monitor their numbers over time. Thermal imaging is used in night counts to census deer populations, allowing managers to count animals from long distances with minimal disturbance to the animals (The Deer Initiative 2008). It has also been used to find injured animals during culling (FLIR 2021).

Poaching

Poaching, the illegal hunting of wildlife species by humans, is a well-known but complex issue (Montgomery 2020). While the scale of this crime varies dramatically on an individual basis in terms of the numbers and status of species affected, the cumulative global impacts of the practice are devastating for wildlife and humans alike. Yet the data to properly understand and address this issue is lacking (Duffy 2016). Many strategies are needed to tackle poaching and the increasing use of technology, including thermal imaging, is among them (Duffy 2016). Hart et al. (2015) trialled the use of the technology in order to evaluate its usefulness as a tool to help wildlife managers combat increasing issues of this all too common issue. They concluded that thermal imaging holds considerable promise for the detection of poachers, but that those considering the tool should take into consideration the effects of habitat types on its detection capabilities.

An iconic species group that has long been a prime target for poaching is rhinoceroses (an example thermal image of a rhinoceros can be seen in Figure 17). Relatively few individuals of the five species of rhino still exist in the wild, and efforts to save them from extinction must include the prevention of poaching. Thermal imaging is among the techniques that have been tested to help tackle the poaching of rhinos in Africa (Mulero-Pazmany 2014). Yet it is not in widespread use: according to a study on the anti-poaching strategies used by private rhino owners in South Africa, only one out of the 22 participants implemented thermal imaging as part of their anti-poaching efforts (Chapman and White 2020). In this study, the participating organisations were anonymised and there were no specific details given on the device or how it was used. An exceptional case is that of an individual blind South-western Black Rhinoceros *Diceros bicornis occidentalis* rescued in South Africa due to its vulnerability and taken to a secure area for protection. A custom security system combining FLIR thermal and visible-light cameras was installed in this animal's enclosure and is now used for added round the clock protection, regardless of light levels (FLIR 2020).

Since trials were carried out by Hart et al. (2015) and Mulero-Pazmany (2014) to investigate this application there have been considerable advances in the available thermal-imaging devices and more are expected. Yet the technology still appears to be underused at present. However, there is enormous potential for the use of thermal-imaging technology as a tool to help prevent poaching in future. Using thermal-imaging drones appears to hold the highest promise here: if used appropriately, they could allow more rapid and effective responses to locate both rhinos and poachers and deploy anti-poaching teams to the areas where they are needed efficiently.

Figure 17 A thermal image of a White Rhino *Ceratotherium simum* ssp. *simum* taken using a FLIR T1030sc thermal-imaging camera.

5.2.2 Agriculture

Agriculture is the second biggest threat to wildlife (Ritchie and Roser 2021; Maxwell et al. 2016). Wildlife habitats around the world are shrinking as agriculture takes up increasing amounts of land in order to provide food for the growing human population. Human–elephant conflict is just one result of the spread of agricultural land-take, and the consequences can be fatal on both sides (Schaffer et al. 2019). Through the Human–Elephant Alert Technologies (HEAT) project, a collaboration between the Arribada Initiative and The Zoological Society of London (ZSL), researchers are developing low-cost thermal-imaging systems to help address this escalating issue (Dangerfield 2020). The team have collected tens of thousands of thermal images of Asian Elephants *Elephas maximus* and African Elephants *Loxodonta africana* in zoos in the UK. They then labelled these images manually to inform the training of a computer algorithm to allow the automatic detection of elephants in the wild. They aim to deploy this in the field as an automated detection system that can alert humans to the presence of elephants (Colchester Zoo 2021). While the initial results of these tests are promising, the images have been taken in what is likely to be very different conditions from those in the field of deployment and so further work in more realistic settings are expected but are yet to be reported on for this project. If successful, projects like this could help us to better understand wildlife behaviours to help avoid direct human–wildlife conflicts and promote human–wildlife coexistence.

In many areas, farming of livestock can result in interactions and conflicts between stock animals and wild predatory animals. This can result in several negative consequences, including stress, injury and death for both stock and wild animals, as well as economic loss for humans. Historically, livestock-guarding dogs (LGDs) have been used by humans around the world in an attempt to protect stock animals from predators (Landry et al. 2020). Despite the use of LGDs in modern farming, in recent years stock loss attributed to returning Grey Wolves *Canis lupus* in the French Alps has increased (Landry et al. 2020). Landry et al. (2020) used thermal imaging to observe night-time interactions between LGDs and wolves in order to investigate LGD behaviour and efficacy. Using SAFRAN MATIS cooled thermal-imaging binoculars (SAFRAN Electronics & Defense, Paris, France) connected to an external recording device, they were able to directly observe and record behaviours. They could then categorise LGD–wolf interactions, as well as documenting wolf attempts to approach or attack stock. The long-range capabilities of the thermal-imaging equipment they used allowed them to observe their target animals non-invasively from station points between 300 and 700 m away from each flock, and they report that they were able to scan the landscape up to 5 km away (though a lack of detailed specifications for their equipment makes it impossible to verify the likely detection range they could achieve). Their results suggest that LGDs may be somewhat effective in protecting stock, with wolves retreating from flocks in 65% of cases. Agonistic behaviours from LGDs were common (e.g. chasing of wolves) and physical fights even occurred in some instances. However, the study also highlighted other more peaceable interactions between LGDs and wolves, including social investigation and instigation of play. The use of thermal imaging allowed the researchers to uncover the complexity of interactions between LGDs and wolves, which would otherwise have been impossible to see using conventional visual methods. There is huge potential for thermal imaging to help provide insights like this at these interfaces between humans and wildlife in a wide range of situations.

5.2.3 Climate change

The effects of climate change threaten wildlife (and humans) in many different ways. More and more it is escalating the effects of other drivers of biodiversity loss, intensifying pressures such as the loss of functional habitat for many species (Mazor et al. 2018). On the front line, Polar Bears' icy habitats are disappearing, forcing them into ever closer contact with humans. This can be dangerous for both humans and Polar Bears, meaning strategies are needed to avoid fatalities on either side. Researchers have been working on low-cost thermal-imaging systems to detect Polar Bears and alert local communities about the presence of Polar Bears in their area (Arribada 2018).

5.2.4 Energy production

Efforts continue to supply energy for ever-increasing global demand, which is predicted to increase by up to 50% in the next few decades (Gordon and Weber 2021), and this can have consequences for wildlife.

Oil and gas exploration is a well-known environmental issue, causing disturbance to wildlife and destruction to their habitats (WWF 2020). An iconic wild animal threatened by exploration activities is the Polar Bear, and although thermal imaging has been used for this species, its efficacy has been problematic due to misuse of the technology (you can read more about this in Section 6.4.3 on Polar Bears).

Wind power is often seen as a much 'greener' method of energy generation and increasing pressure for renewable energy has meant harnessing the wind to generate electricity has grown in popularity. Unfortunately, however, wind power technologies have ecological costs that can go unnoticed (Wright et al. 2020). Researchers identified that bats (Horn et al. 2008; Cryan et al. 2014) and birds (Stewart, Pullin and Coles 2007) are at risk from wind energy generation sites. Thermal imaging has helped to make these discoveries at both onshore (Horn et al. 2008; Cryan et al. 2014) and offshore wind energy facilities (Cullinan et al. 2015; Matzner et al. 2015). Since then, thermal imaging has also been used to investigate ways to address the risks to volant wildlife (Kinzie 2018, Gilmour 2019, Schirmacher 2020, Smallwood 2020).

The relatively long lifespans and low reproductive rates of bats and some birds (Barclay and Harder 2003; Munshi-South and Wilkinson 2010) make them especially vulnerable to negative impacts from wind turbines. These impacts can come in a variety of forms, including collision or barotrauma (direct impacts), displacement, disturbance, and behavioural changes (indirect impacts), barrier effects and habitat degradation (Baerwald et al. 2008; Goodale and Milman 2014; Kaldellis et al. 2016). One approach to alleviating these pressures on bats at wind turbines is to use acoustic deterrents, which broadcast ultrasound to repel bats. Testing the efficacy of these devices using conventional acoustic techniques can be problematic because the output of the deterrents being tested can interfere with acoustic recordings. For this reason, thermal imaging has been used to help assess bat responses to these deterrents, as well as to obtain spatial and behavioural data (Kinzie et al. 2018; Gilmour 2019; Schirmacher 2020). The use of multi-camera systems to gain 3D positioning of bats (sometimes referred to as stereo thermal videogrammetry) has enabled investigators to reconstruct bat flight paths as visualisations, providing a greater understanding of bat movements in response to the deterrent strategies being tested (Kinzie et al. 2018; Gilmour 2019; Schirmacher 2020). Deploying systems like these are not easy, however, as they can involve complex systems, use of specialist software and also require target or wand-based calibration of space, or alternative software procedures (Betke et al.

2017; Gilmour 2019) to help with correct positioning. In spite of these challenges, the resulting data can be invaluable and these methods have revealed useful information to quantify the likely effectiveness and functional spatial range of deterrents (Kinzie et al. 2018; Gilmour 2019; Schirmacher 2020). Another approach to reducing fatalities of both birds and bats at wind turbines is to periodically curtail their operation, which Smallwood and Bell (2020) were able to study using thermal imaging and detection dogs. In another related study Smallwood et al. (2020) used thermal imaging to identify periods of peak bat activity which allowed them to focus their detection dog operations more efficiently to find carcasses at wind turbine sites. One of the main issues with using thermal imaging for wind energy projects has been the vast amount of data they can generate, which must then be analysed and reported on. Thankfully, automated software is now available to help to detect and track bats in thermal video (Corcoran, A. J. et al. 2021).

Solar power tower facilities generate electricity using thermal energy from the sun. They are made up of a complex system of sun-tracking mirrors (called heliostats) that reflect solar energy to a focal receiver onto a central tower, which can then be passed to other system components to be converted and stored for later use. In a pilot study carried out in California in the USA, Diehl et al. (2016) evaluated five different video-based monitoring technologies (including two different thermal-imaging devices) and radar to detect wildlife at a solar power tower facility. These facilities present challenging field conditions where target species groups for monitoring include insects, bats and birds, which may be at risk of mortality (combustion) there. The first thermal camera they used was an Axis Model Q1921-E with 19 mm and 35 mm lenses (Axis Communications, Lund, Sweden) which they referred to as a *thermal surveillance* (TS) camera. The second was a FLIR SC8343HD (FLIR Systems, Billerica, Massachusetts, USA) with a telephoto lens which they referred to as a *scientific-grade thermal camera* (SGT). They found that the TS worked best in combination with a surveillance camera operating in the visible spectrum. This combination of cameras was favoured as a monitoring solution due to their ready availability and practical suitability for unattended longer term field use. They reported that the SGT on the other hand, while producing quality imagery, was of limited practicality for unattended longer term monitoring due to the unfeasible volumes of data it generated and the value of the equipment.

5.2.5 Transport and infrastructure

Wildlife vehicle collisions (WVCs) have huge implications for wildlife and humans: they can pose a serious threat to wildlife populations and for humans, they can cause major issues for safety, traffic management and also have serious economic costs (Faber Johannesen 2021). In the UK alone, 15,631 wild animals killed on roads were reported in 2019 (Project Splatter 2020), while insurance claim data suggests almost 32,000 WVCs occur each year resulting in around £63.8 million in insurance payouts (Zurich 2021). In an attempt to tackle this global problem, many have devised mitigation strategies which have been met with varying levels of success (D'Angelo et al. 2010; McCollister and Van Manen 2010).

Mitigation approaches include the use of wildlife tunnels, underpasses and bridges to allow animals to traverse infrastructure such as roads and railways (Smith et al. 2015; Mysłajek et al. 2020; Gregory et al. 2021). In a UK-based study, Faber Johannesen (2018) evaluated the use of thermal-imaging, camera-trap and sandtrap methods to monitor wildlife tunnels designed as part of a mitigation scheme. These tunnels were created during the construction of major roads to allow the safe passage of wildlife underneath them. Faber

Johannesen's (2018) wildlife tunnel research showed that thermal imaging facilitated the detection of a higher diversity of wildlife species and a greater number of individual animals overall when compared to camera trapping or sand trapping data. Not only did they find that thermal imaging detected more wildlife, but also that the data obtained provided a greater breadth of information on the behaviour of these animals in relation to the infrastructure and surrounding habitat. Faber Johannesen (2018) recommends that thermal imaging should be used for wildlife monitoring on large-scale infrastructure projects.

Aviation is another area of human activity that can have an impact on wildlife, particularly birds. Bird strike can have detrimental impacts for both aviation and avian safety (Metz et al. 2020) giving rise to $1.2 billion in annual financial losses (Medolago et al. 2021). In order to tackle this issue, avian radars dedicated to tracking birds have been installed at airports and are increasing in popularity (Metz et al. 2020). Metz et al. (2020) suggested that using thermal imaging in combination with these avian radar systems has the potential to provide real-time bird strike prevention. Medolago et al. (2021), however, trialled the use of a FLIR T650sc thermal-imaging camera as a stand-alone approach. They used this device on walked transects as part of their study of technology to detect birds at an airport (you can read more about this study in Section 3.3.2).

5.3 Behavioural studies

Understanding the behaviour of a species can be a critical component of the success of efforts to manage or protect them (Greggor et al. 2019). Behavioural studies can help to provide useful data to inform practical decisions and actions made by wildlife professionals and relevant stakeholders. Using thermal imaging is a non-invasive approach that can allow observations in all light levels with minimal disturbance, minimising potentially detrimental effects on behaviours in focus and the implications they may have for data outputs.

Miard (2020) used thermal imaging, in combination with red filtered head torches, to study the behaviour of the Sunda Colugo *Galeopterus variegatus* in the wild on Langkawi Island, Malaysia. Colugos are strictly arboreal mammals, closely related to primates. These creatures are nocturnal and well camouflaged against the trees they live in. Taken together, the ecology of these fascinating mammals makes their behaviour almost impossible to study using traditional direct observational methods. However, using a handheld FLIR III 640 thermal-imaging scope and a torch with a red-light filter (Clulite HL13), Miard (2020) was able to observe and document the behaviour of these animals. This allowed them to construct ethograms documenting 19 different behaviours over 265 hours of observation from all 24 hours of the day. They were also able to establish activity budgets for colugos, which showed onset of activity within an hour after sunset, with a brief peak rest period between midnight and 01:00, ending with animals returning to their resting sites around sunrise. This research also revealed that, in contrast to previous theories, colugos live in social groups within discrete territories. It is important to note that colugos and other nocturnal mammals are sensitive to light, which can cause disturbance and eyesight problems. Therefore using non-invasive methods like thermal imaging can be crucial to gaining useful and valid information on their natural behaviour in the wild.

Studying avoidance behaviours of birds in relation to light, Syposz et al. (2021) conducted field experiments on the island of Skomer in the UK. They used a FLIR T620 thermal-imaging camera to record the behavioural responses of Manx Shearwater *Puffinus puffinus* at a breeding colony on the island to different light stimuli at night. The use of a non-invasive, visual method of recording was critical to gaining the data Syposz et al. (2021) needed to examine how these birds reacted to a variety of experimental light conditions. To create these conditions, they used a torch (Icefire T50) and gel filters, making five different light condition treatments: blue, green, red, dimmed white and bright white light. They also varied the duration of the periods these lights were switched on. Using an automated analysis procedure in MATLAB, they were able to detect, track and count the two-dimensional flight paths of individual birds under each condition. Syposz et al. (2021) found that adult Manx Shearwater avoided the artificial light stimuli presented. However, these responses were not equal: red light caused the least disturbance, whereas shorter wavelengths of light (blue and green) and higher brightness caused higher levels of disturbance.

Anti-predator behaviours are crucial for wildlife survival, and they can be particularly important at night when rest or sleep may make animals particularly vulnerable. Yet studying such behaviours at night can be intrinsically onerous. In a study carried out in Namibia, Burger et al. (2020) investigated the nightly behaviours of the Angolan Giraffe *Giraffa giraffa angolensis*. Their use of multiple thermal-imaging systems in combination with a single night vision system gave the researchers a fascinating insight into the rest, sleep and anti-predator behaviours of these enormous African mammals. They used SEEK CompactPro (Seek Thermal, Inc., CA, USA) thermal-imaging devices connected to iPad tablets (Apple Inc., CA, USA), which helped to reveal sleep guarding behaviours displayed by the giraffes during hours of darkness; something they could not otherwise have seen without the use of these technologies. Giraffes were found to gather in groups at resting sites around sunset. The location of these resting sites changed nightly, and by using vehicle-based systems the research team were able to follow groups to each site and then set up a stationary sampling point at a safe distance from the animals. There they observed that giraffes in the group would rest and lay down to sleep, but at least one giraffe would remain awake. These individuals seemed to act as sentinels, staying watchful and actively scanning their surroundings.

Understanding migration behaviour is vital to the conservation of migratory species (Horns and Şekercioğlu 2018; Gao et al. 2020). Studies using thermal imaging to investigate the migratory behaviours of wildlife have so far focused primarily on birds (Anderson et al. 2021, Fortin et al. 1999, Gauthreaux and Livingston 2006, Liechti et al. 1995, Liechti et al. 2003, Zehnder et al. 2001) and these are covered in Section 6.2.

5.4 Health and rehabilitation

5.4.1 Health

Thermal imaging has been demonstrated as a useful tool to aid veterinary applications for domestic, agricultural and wild animals (Cilulko 2013; McCafferty 2007; Hilsberg-Merz 2008). Notably, it has been employed as a complementary diagnostic tool in equine veterinary medicine since the 1970s (Soroko and Morel 2016) and to look for signs of disease in the animal production industry (Naas 2014).

Thermal imaging has been used to facilitate the detection of a range of animal health conditions including:

- ocular disorders (Biondi et al. 2015)
- spine-related issues (Graf von Schweinitz 1999)
- respiratory disease (Schaefer 2012)
- inflammatory disorders (Graciano et al. 2014)
- lameness (Eddy et al. 2001)

- infection (Islam 2015)
- mastitis (Polat 2010)

One of the advantages reported is that the sensitivity of the technique allows for early detection of some conditions or the detection of subclinical differences that might otherwise go unnoticed using other methods (Islam 2015; Polat 2010). Thermal imaging has also been used to detect estrus (Sykes 2012) and pregnancy in domestic and wild animal species (Domino et al. 2021; Hilsberg-Merz 2008).

Established procedures for effective thermal imaging for animal health applications involve the use of science or clinical grade equipment by appropriate personnel under appropriate conditions (Hilsberg-Merz 2008). Veterinary diagnosis should only be carried out by a qualified veterinary professional. However, images to aid in diagnosis can be collected by trained imaging technicians. In practice, unfortunately, there are many individuals without the relevant qualifications or expertise who are offering thermal scanning services. These are particularly common in the equine sector, but such malpractices have the potential to be easily transferred to wild and feral equids and a wide range of other wildlife species. To prevent this, mechanisms to regulate the use of thermal-imaging technology for diagnostic purposes are needed.

5.4.2 Rehabilitation

In a fascinating and well-conducted study on the use of thermal imaging to assess Hedgehogs *Erinaceus europaeus* admitted for rehabilitation at a rescue centre in the UK, South et al. (2020) found that the use of this technology provided significant benefits in their work. This allowed them to more accurately assess Hedgehogs for hypothermia, which has significant implications for their health, welfare and survival. When Hedgehogs are rescued, it is rarely possible to obtain core temperature values, which makes hypothermia assessments difficult to carry out. As Hedgehogs are the most common mammal admitted to wildlife hospitals and rescue centres in the UK, having a rapid, non-invasive and accurate way to assess them could save time, money and most importantly numbers of surviving members of this declining species.

South et al. (2020) compared thermal-imaging techniques with conventional observational assessment by centre staff to classify the hypothermic status of Hedgehogs on admission. For the former part, they took facial thermal images in a controlled environment using a FLIR E60bx. From these images, they calculated temperature based on the apparent corneal temperature, which was duly corrected by accounting for the ambient temperature, reflected temperature, distance and emissivity. The eye region is commonly used as a proxy for core body temperature in animals (Giannetto 2021), but South et al. (2020) first verified that the relationship of the corneal reading parameter was within 0.1 °C of the core body temperature of Hedgehogs prior to using it in their hypothermia assessments. If transferrable to other wild animals, which requires further research, this could mean that conventional invasive rectal temperatures could potentially be replaced by the non-invasive use of thermal imaging as a proxy for core body temperature.

Thermal imaging could also be used as a tool to help rehabilitators find sick or injured animals in the field. This could potentially save considerable time and lives in situations where animals may be difficult to locate due to light levels or other challenges that might impede swift visual detection.

Thermal imaging is not thought to be used in rehabilitation currently, yet it could have huge benefits if properly researched, developed and applied effectively.

5.5 Thermoregulation, thermal biology and thermal ecology

Animals can regulate their body temperature through physiological, behavioural and physical processes (Tattersall and Cadena 2010). Thermal-imaging technology has been instrumental in a range of exciting discoveries in the fields of thermoregulation, thermal biology and thermal ecology, among others, that focus on these processes. The non-invasive nature of thermal-imaging technology allows researchers to take thermal readings without disturbing the processes they are studying (Tattersall 2016).

One area where thermal imaging has been particularly instrumental in visually demonstrating a thermoregulatory phenomenon is the study of thermal windows. Many animals use these specialised regions of their body surface to help regulate their temperature in changing thermal conditions (Andrade 2015). Such functional areas of heat exchange come in a range of forms and some have been identified with the help of thermal-imaging technology, including:

- ears of African Elephants *Loxodonta africana* (Phillips and Heath 1992)
- bills of Toco Toucan *Ramphastos toco* (Tattersall, Andrade and Abe 2009)
- features of the head, lower legs and paws of foxes (Klir and Heath 1992)

These thermal windows allow animals to adapt to variations in ambient temperature by regulating heat transfer between their bodies and immediate surroundings. Like windows, they can be opened and closed as required (Tattersall 2016).

5.6 Communication, education and awareness

The use of visual media can be a powerful tool to communicate important information to a wide range of audiences. Thermal images can be striking and can help to draw attention to a subject or species that might have otherwise gone unnoticed. In any wildlife-related discipline, we can use these images to help us to communicate important information, educate others and raise awareness about important issues. At a very basic level, this may be as simple as using a thermal image in a report, paper or article to illustrate a particular finding, concept or result. Clips of thermal video footage can be used in presentations to clearly show particular behaviours or phenomena to an audience.

Of course, thermal-imaging devices and systems can also be used in real-time out in the field to help communicate, educate and raise awareness too. Seeing the technology in action often interests a diverse range of people, this can help to engage individuals who might not otherwise have an interest in wildlife *per se*. Reaching a wider audience and spreading awareness is critical in addressing the growing number of wildlife-related issues we face, so it is well worth considering how we might incorporate these technologies into activities and projects that involve the wider community.

In mainstream media, thermal imaging is used in wildlife television programmes and documentaries. On television in the UK, *Winterwatch* (BBC 2015) showed thermal-imaging footage of Grey Seals *Halichoerus grypus* at Blakeney Point in Norfolk. The BBC continues to feature the use of this technology intermittently in similar programmes, including *Into the Bat Cave* (BBC 2019), which used the Leonardo Merlin and Horizon thermal-imaging cameras (Leonardo, UK Ltd, London, UK) to film Greater Horseshoe Bats *Rhinolophus ferrumequinum*. In *Planet Earth II* (BBC 2016), the Leonardo SLX Hawk thermal-imaging cameras (Leonardo, UK Ltd, London, UK) were used to film Leopards *Panthera pardus fusca* in Mumbai. The documentary series *Night On Earth* (Netflix 2020) used a range of different

Case study: 'Night Spotting' with Dr Priscillia Miard

Focusing on the nocturnal mammals of South and Southeast Asia, the Night Spotting Project (NSP) is an outreach research project which uses a range of technologies, including thermal imaging, to study:

- Sunda Colugo *Galeopterus variegatus*
- Sunda Slow Loris *Nycticebus coucang*
- Red Giant Flying Squirrel *Petaurista petaurista*

Dr Priscillia Miard is the Founder and Lead Researcher of the (NSP). Through NSP, Priscillia has helped to gather valuable information on these species and to engage a wider range of people to learn more about them.

Why do they use thermal imaging?

Sensitive, secretive and nocturnal species like these can be tricky to find. Colugos in particular are well camouflaged against the tree trunks they rest on, and cannot be detected easily by the eyeshine that is often used for the detection of other species. NSP's focal species are understudied and often go unnoticed by humans, yet they provide vital ecosystem services, including pollination, pest control and tree productivity. It is important that researchers use the right tools to gain the data that is required to better understand and protect these species. Priscillia's PhD research on these animals revealed that using thermal imaging alongside a torch with a red-light filter gave the best results in terms of detection and avoided issues of disturbance and blinding of the wildlife too.

How do they use it?

Priscillia and the NSP team use handheld thermal-imaging scopes (including FLIR Scout III model 640) as an aid to help them spot their target species in the wild. They conduct transect surveys on foot, with the help of red-light torches to guide them along their way, scanning the surrounding vegetation regularly using their thermal-imaging device for the direct detection of their target species.

What benefits has it given them?

Thermal imaging has provided several benefits to Priscillia and the NSP team. First, it is **non-invasive**, which allows them to study their target species with minimal disturbance. Second, it allows them to detect their target species no matter how light or dark it is, and to show it to others in **real-time**. Finally, the handheld device they use is lightweight, portable and durable, making it ideal for **practical** use in the field.

For more information on the Night Spotting Project (NSP) visit their Facebook Page at: https://www.facebook.com/NightSpottingProject/

camera technologies to capture the nocturnal behaviour of wildlife, including thermal-imaging footage of Wild Guanacos *Lama guanicoe* in Chile captured using a drone-mounted camera from FLIR's T1K Series (FLIR Systems, Inc., Wilsonville, OR, USA). Few thermal-imaging cameras are capable of attaining the imagery required for television-quality production. Most of the thermal-imaging footage seen on our screens through traditional media has been shot using high specification cameras. These are often military-grade or custom-made equipment allowing high-resolution footage that is far beyond what can be achieved using even most scientific-grade thermal equipment.

Wildlife photographers are now also using thermal imaging to help them detect wildlife for them to capture using their non-thermal cameras (A. Rowley pers. comm. 2021). As with any detection based application, this can help to save considerable amounts of time and effort, particularly for cryptic, secretive or otherwise tricky species to photograph.

6. Wildlife Applications

In the following sections, we will explore the key wildlife groups that have been the focus of work in the field of thermal imaging for wildlife to date. From the literature reviewed on the topic between 1968 and 2021 (inclusive), these key functional application groups include:

- mammals
- birds
- bats
- marine mammals
- marsupials
- insects
- fish

The number of publications differs markedly between these groups (see Figure 18). We will discuss some of the main research and information available for the top four of them. In each case, we will not only review the key literature itself, but we consider how the collective results can be applied from a practical perspective. As the majority of species applications require detection in some way, we consider the results primarily in this context.

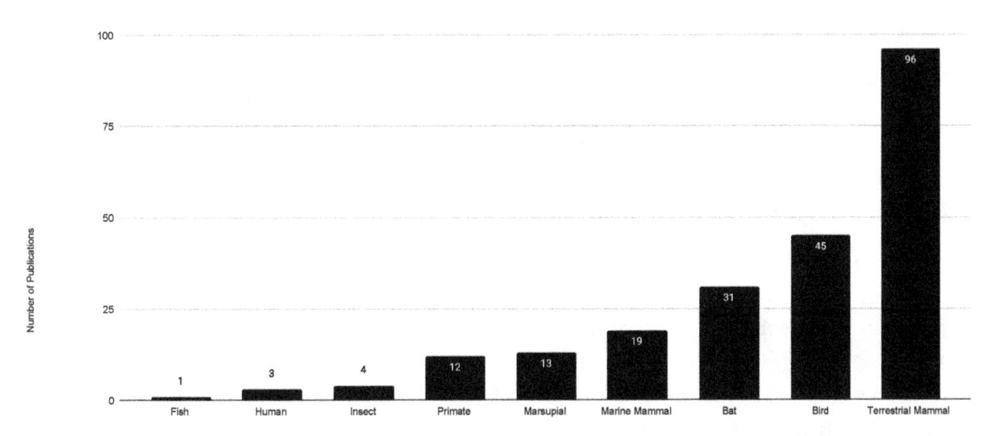

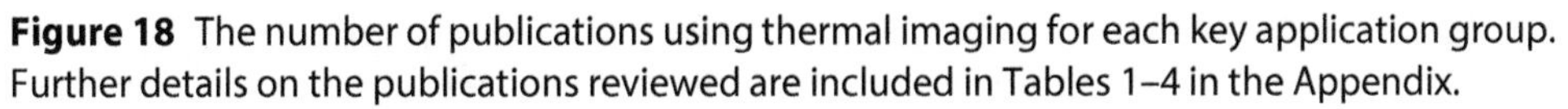

Figure 18 The number of publications using thermal imaging for each key application group. Further details on the publications reviewed are included in Tables 1–4 in the Appendix.

6.1 Mammals

Nocturnality is the most common activity pattern observed (44%) in mammals, followed by crepuscular and cathemeral behaviours (29%). Only 26% of mammals are diurnal (Jones et al. 2009; Miard 2020). The majority of mammals will, therefore, present challenges to detection due having activity patterns that are not conducive to traditional direct observation methods (due to variable or low light levels). It is not surprising that mammals, as a group, are the most studied using thermal imaging, where it has the potential to help improve survey accuracy, especially for the large proportion of species that are active during hours of darkness or low light levels.

As the most studied group, turning up 111 accessible publications found between 1968 and 2021, this section highlights the key findings in the literature and is not intended as a comprehensive review of work to date. However, you can find a more comprehensive list with more details and key information from the literature including the use of thermal imaging for mammals in Table 2 in the Appendix.

6.1.1 Deer

As discussed earlier in this book, deer have been the most popular subject of thermal-imaging studies and were the first to be the focus of thermal-imaging detection research (Croon 1968; Graves 1972; Parker 1972; Wride 1977). The earliest of these used aerial survey techniques, where primitive thermal-imaging devices were used from fixed-wing aircraft. Ground-based approaches have since been used (e.g. Kilgo et al. 2020) and more recently researchers have also taken to surveying from the air using drones instead of aeroplanes.

In a US-based study Beaver et al. (2020) tested the use of thermal drones for deer at Auburn's deer research facility in Alabama. This provided a controlled but natural setting with known numbers of deer present within a set boundary. Using a FLIR Vue Pro 640 (resolution 640 × 480 pixels) mounted on a Ritewing Drak fixed-wing drone they flew at a height of 100 m AGL (above ground level) in a transect grid pattern using a semi-automated data capture process. They analysed their data manually to count deer from their thermal-imaging footage. When they sampled under appropriate conditions (in this case by timing surveys in the evening) they were able to detect 92% of known individual deer in their sample area.

Deer management is a key application for thermal imaging. In the UK, ground- or vehicle-based thermal imaging is now in widespread use as an aid for deer managers. In the USA, Fairfax County Deer Management Program have deployed aerial and ground-based thermal-imaging techniques to aid their deer management operations (Hodnett 2005).

6.1.2 Primates and arboreal mammals

Primate studies began on chimpanzees in captivity (Kano 2016), and the first study on orang-utans in the wild in 2018 (Dahlen 2018). Since then, several studies have used thermal technology to detect this group of mammals (Ampeng 2021; Rahman 2020; Jumail 2020; Spaan 2019; Semel 2020; Kays 2019; Heintz 2019; Burke 2019; Abdul-Mutalib 2019; Dahlen 2018; Dezecache 2017; Kano 2016). Burke et al. (2019) tested the use of a drone with a dual camera system to detect Bornean Orang-utans *Pongo pygmaeus* and Proboscis Monkeys *Nasalis larvatus* in the wild in Sabah, Borneo, Malaysia. Their equipment consisted of a FLIR Tau 2 640 thermal-imaging camera core (640 × 512 pixel resolution and 19 mm lens) alongside a TAMRON visible spectrum camera (1920 × 1080 pixel resolution) mounted on a Tarot X4 drone. The researchers conducted thorough testing of their set-up before they began sampling, allowing them to establish a workable protocol for their main fieldwork for this study. They studied wild Orang-utans in two ways: (1) by flying the drone over known nest locations and (2) by conducting a 'blind' flight in a grid pattern over habitat where nest locations were unknown. They also trialled flying the drone to detect Proboscis Monkeys at known locations (3). In scenario (1) thermal imaging allowed them to detect all 28 of the Orang-utans that had been located in advance by tracking teams. In scenario (2) thermal imaging revealed 13 Orang-utans that were then verified by ground teams, showing no false positives, though false negatives could not be quantified. Lastly, scenario (3) showed a troop of 11 Proboscis Monkeys, which were confirmed by ground teams to identify the species, although it was not possible to count the number of individuals to confirm if there actually were 11. The authors conclude that thermal drones can be used to detect primates in tropical rainforest habitats, although further refinement may be needed to gain better spatial resolution and species identification.

Miard (2020) evaluated the use of thermal-imaging, red-light and white-light methods for the study of arboreal mammals in Malaysia. They used a handheld FLIR Scout III 640 thermal-imaging scope as a ground-based method for detecting a range of species including: Sunda Colugo *Galeopterus variegatus*, Sunda Slow Loris *Nycticebus coucang* and Red Giant Flying Squirrel *Petaurista petaurista*. Miard (2020) found that using thermal imaging and red light significantly enhanced their wildlife detection compared to the use of a white-light head torch (Clulite HL13). This improved detection allowed them to find more species and more individual animals. However, it was not just the detection rate that saw an improvement: using this method is also better for the wildlife too. White light is still commonly used by wildlife researchers in the field, but it can be detrimental to both target and non-target species. First, white light can cause disturbance, which may change both the short and long-term behaviour of the affected animals. Second, nocturnal mammals can be blinded by white light. Using alternative non-invasive methods such as thermal imaging in combination with red light (where required) could help to improve data on a wider range of species, while also decreasing the impact on the eyesight, behaviour and ultimate survival of the animals themselves.

In a similar approach to Miard (2020), Zainalabidin et al. (2020) used thermal imaging in combination with a red-light filtered head torch to gather baseline data on the distribution of five nocturnal arboreal mammals in Borneo: Bornean Striped Palm Civet *Arctogalidia stigmatica*, Philippine Slow Loris *Nycticebus menagensis*, Western Tarsier *Cephalophacus bancanus*, Bornean Colugo *Galeopterus borneanus* and Island Palm Civet *Paradoxurus philippinensis*. Using walked transects and point count methods, researchers used a handheld FLIR III 640 thermal-imaging scope to scan for their target species from ground level. Using this technique, the researchers were able to detect all five of their target mammal species as well as a further 20 non-target mammal species in their study areas.

6.1.3 Carnivores

Bushaw et al. (2019) reported that by using a drone-based thermal-imaging system they could reliably detect mesocarnivores in agricultural and grassland habitats in Manitoba, Canada. Using a DJI Zenmuse XT2 R thermal-imaging camera mounted on a DJI Inspire 1 drone (Da-Jiang Innovations, Shenzen, Guangdong, China), they conducted point counts using pre-programmed flight routes using ArcGIS and the DJI Ground Station Pro v 2.1

App. The target species group for this study included: Coyote *Canis latrans*, Red Fox *Vulpes vulpes* and Striped Skunk *Mephitis mephitis* which they encountered in low numbers. As well as detecting these, they were also able to pick up smaller, non-target, animals such as mice and birds, which assured them that had their larger target species been present in larger numbers, they would have been able to detect them easily.

Brawata et al. (2013) used thermal imaging to study carnivore behaviour which you can read more on in Section 3.3.1.

6.1.4 Small to medium-sized mammals

McGregor et al. (2021) attempted a comparison of thermal-imaging versus spotlighting methods to find small to medium-sized mammals in arid habitats in Australia. They used a FLIR III 640 thermal-imaging scope operated from the rear passenger seat of a vehicle, while another person operated a spotlight from the front passenger window, with both devices facing in the same direction. Unfortunately, it appears from their methods that their data had a high potential for bias: the two operators of each technology were allowed to communicate, which could result in informing one or other of animals they may not otherwise have seen. This said, they reported that when they used thermal imaging in appropriate conditions thermal imaging provided a 30–50% increase in small mammal detection rates.

Lagomorphs

Lagomorphs can be difficult to detect, often because they use a range of mechanisms to evade predators, including humans. Numbers of Brown Hare *Lepus europaeus* have declined across Europe and several studies have used thermal imaging to detect them with this fact in mind (Voigt 2020; Karp 2020; Sliwinski et al. 2021). Karp (2020) and Sliwinski et al. (2021) both used vehicle-mounted thermal-imaging devices to detect Brown Hares (you can read more details about these in Section 3.3.6). Voigt and Siebert (2020) deployed thermal imaging using vehicle-mounted and ground-based transect methods. These researchers needed to detect leverets in an agricultural landscape in Germany in order to examine the survival rates of Brown Hares. Finding these animals is inherently difficult as leverets, by their nature, are small, inconspicuous and stay very still for long periods of time. The device used was a Raytheon Palm IR 250-D (320 × 240 pixel resolution), and although it was reported that this device allowed for reliable detection, identification was difficult due to its low resolution. In such situations, they had to approach the detected animal more closely and inspect it with a torch for identification purposes. Such disturbances could have potentially been negated by using a device with more suitable specifications for this task. Bedson (2021) tested the use of thermal imaging, among other technologies, to detect another often elusive lagomorph species: the Mountain Hare *Lepus timidus*. This species was driven to extinction in the UK thousands of years ago, but was reintroduced to Holme Moss, over 150 years ago where Bedson's (2021) study took place. Using an Armasight Command 336 HD, a thermal scope with binocular output (Armasight Inc., San Francisco, CA, USA), they were able to detect hares up to 740 m away. During night-time surveys they used a point transect method, where observers walked set transects and then stopped at key sampling points, set up the thermal-imaging device on a tripod and conducted 360° scans of the area. Psiroukis (2021) took a drone-based approach to use thermal imaging to detect European Wild Rabbits *Oryctolagus cuniculus* on the island of Lemnos, Greece. For this study they used a FLIR Vue Pro thermal-imaging camera (336 × 256 pixel resolution) mounted on a DJI Phantom 4 Pro drone to conduct pre-planned night-time flights. Before they began their sampling for this study, they conducted test flights using captive rabbits which allowed them to determine an appropriate flight altitude (25 m above ground level) to maximise detection and minimise disturbance.

Hedgehogs

The European Hedgehog *Erinaceus europaeus* can be difficult to locate due to its small size and nocturnal behaviour, but with populations in considerable decline in the UK it is important to use techniques that can provide accurate data on numbers. Bowen et al. (2020) and Gurnell et al. (2021) report on the successful use of thermal imaging as part of monitoring surveys carried out in Regent's Park, London, UK: the location of the only breeding population of Hedgehogs in the city. Using a systematic approach, trained staff and volunteers used a FLIR E60 thermal-imaging camera (FLIR Systems UK, West Malling, Kent). Alongside more traditional spotlighting methods, this handheld thermal-imaging device was used during transect surveys where surveyors walked and stopped at regular intervals to scan their surroundings for Hedgehogs. They found that thermal imaging was particularly useful as it allowed them to pick up animals at greater distances than they could when using other techniques. It should be noted that the habitats were largely open and without dense vegetation, which allowed a clear line of sight – this should, naturally, make detection easier. However, the key reasons for the success of this programme likely also include the fact that thermal imaging has not been used blindly here: the team liaised with experts in the field of thermal imaging to develop a standard protocol and then provided device-specific training to surveyors before sampling. Hedgehogs are not always desirable to conservationists, however, and these small mammals were introduced by humans into New Zealand. Considered a pest and an invasive species, it was unknown as to how they might impact local ecosystems. In order to find out more about this, Nottingham et al. (2019) investigated the diets of Hedgehogs in urban forest fragments. To find the Hedgehogs for this study, they used a handheld Pulsar Quantum XD19 thermal-imaging scope (Pulsar, Yukon Advanced Optics Worldwide, Vilnius, Lithuania). Animals were caught by hand and euthanised, and their stomach contents examined. The results suggested that the large quantity of invertebrates they consume may have a significant impact, though further study was required to better understand the potential effects these animals may have on wider ecosystems.

6.2 Birds

Birds were the second wildlife group to be studied using thermal imaging, with papers appearing on the subject in the 1980s and 1990s (Best 1982; Boonstra 1993; Sidle 1993; Garner 1995; Liechti 1995; Benshemesh and Emison 1996; McCafferty 1998; Fortin 1999). This group is also the second largest in terms of their frequency in the literature, of which 45 documents were found and could be accessed during the research for this book (see Table 4 in the Appendix for more details on these). Some studies focus on detecting ground nesting or roosting sites, some focus on birds on or around water or wetland vegetation, while others seek birds in flight.

6.2.1 Ground nesting

One of the most interesting subjects in the early literature for this group is the use of thermal imaging to detect mounds constructed by Malleefowl *Leipoa ocellata* (Benshemesh and Emison 1996). These fascinating Australian birds build large (3–5 m) mounds of earth and vegetation which produce heat via microbial decomposition to incubate their eggs. The mounds are periodically opened by the birds in order to maintain appropriate incubation temperatures, among other possible reasons. The researchers in this study hypothesise that the opening of mounds may present an opportunity for detection using thermal imaging. They used two different thermal-imaging devices and each was deployed differently for different reasons. The first device was a Daedalus Scanner 1240/60 which was mounted on a small aircraft to detect mounds from the air, to test the potential field technique for future use. The second was an Inframetrics 445 mounted on a tower above test mounds, used to more closely examine thermal factors and other variables at play. Analysis was conducted by manual examination of 'quick prints' and video cassette recordings generated from the thermal-imaging device outputs. While the authors declare the method to be feasible, further work would be necessary to develop it into something applicable for practical monitoring purposes. Given how much thermal-imaging technology has progressed since this paper was published, it seems likely that the use of an updated technique using modern technology and relevant expertise could potentially yield much better results for this particular application in future.

Much more recently, in Finland, Santangeli et al. (2020) tested a semi-automated drone-based approach in combination with artificial intelligence to locate the nests of Northern Lapwing *Vanellus vanellus*. These ground-nesting birds are threatened mainly by growing intensification of agriculture, including nest destruction during farming operations. Protecting these nests can be troublesome due to the level of time and difficulty in locating them. As is the case for any wildlife species: if we can't see them, we can't protect them. Santangeli et al. (2020) used a FLIR Tau 2 (336 × 256 pixel resolution) mounted on a DJI Phantom 3 drone and used a ThermalCapture unit (TeAx Technology GmbH, Germany) to allow them to record imagery directly onto a USB stick. They then trained a neural network based on data with known presence or absence of nests. Once trained, they were then able to use this to automate detection of nests from their field data.

Detecting and locating nests can be challenging, but finding European Nightjar *Caprimulgus europaeus* is a particularly onerous task and the usual methods for carrying out these tasks, such as observation, capture and radio tracking, require considerable effort. Shewring and Vafidis (2021) tested an alternative approach to finding this cryptic species using drones with on-board thermal-imaging cameras. Their study was conducted at two forestry sites (upland clear-fell) South Wales, UK: one with known nightjar nesting sites (1) and the second where it was not known if nightjar were present (2). They also used two different drones and thermal-imaging device set-ups. The first was the Falcon 8 drone (Ascending Technologies Ltd.) combined with a Tau 640 thermal-imaging camera with 19 mm lens (640 × 512 pixel resolution). The second was a T600 Inspire 1 drone (DJI Technology Company) with a Zenmuse XT V2.0 FLIR thermal-imaging camera with 19 mm lens (640 × 512 pixel resolution). Unfortunately, the specification of the thermal device used in this study was not the best fit for this application, meaning that low level flight was required for detection: when flying over known nests at site (1), they were able to detect nests around dusk and dawn from altitudes of up to 25 m, but 12–20 m provided their best results. Flying at these altitudes can be risky. With a better match of equipment specifications, in theory, this application could be much more viable if repeated in future.

Fletcher and Baines (2020) used thermal imaging and radiotelemetry to locate Capercaillie *Tetrao urogallus* in their study which investigated the poor breeding success of this large ground-nesting bird. Initially, the researchers used radio telemetry techniques to determine the rough area where tagged female birds were located. They pinpointed the birds exact locations by scanning these areas using a handheld Pulsar Quantum XP50 thermal-imaging scope. This is a good example of using complementary techniques in a two stage approach (see more on combining techniques in Section 3.4). Only two of the 12 breeding attempts they followed were successful (resulting in fledged young).

6.2.2 Water and wetland habitats

Bushaw et al. (2021) used thermal-imaging drones to conduct daytime surveys for broods of dabbling ducks, including Blue-winged Teal *Spatula discors*, Mallard *Anas platyrhynchos*, Northern Shoveler *Spatula clypeata*, Gadwall *Mareca strepera* and Wood Duck *Aix sponsa*. They used a FLIR Zenmuse XT thermal-imaging camera (resolution of 640 × 512 pixels) alongside a DJI Zenmuse X4S visible spectrum optical camera on a DJI Matrice 210 quadcopter drone (Da-Jiang Innovations, Shenzen, Guangdong, China). The thermal camera was used for detection of animals, while the camera operating in the visible spectrum was used for identification of broods. This equipment was used at two different sites with wetland habitats: one in Manitoba, Canada and another in Minnesota, USA. In this study, the use of thermal-imaging drones doubled the detection rates compared to traditional surveying techniques for their target species. The researchers reported dramatic improvements in their ability to detect broods in emergent vegetation, which would not otherwise have been seen at all. They also mentioned that they simply would not have been able to manually detect many of the broods found using thermal imaging, which was likely due to their challenging field conditions and the natural limitations of human perception. The technique also made surveys three times faster than usual, meaning this was not only more effective for detection, but that it was also more efficient too.

Austin et al. (2016) identified that traditional methods for waterbird species were biased towards daytime surveys and that this bias may have serious negative consequences for their conservation. In an Australia-based study, they further investigated this issue by conducting daytime and night-time waterbird counts and comparing the results to determine whether water numbers and/or activity differed. They used binoculars and telescopes to detect waterbirds during the day and a tripod-mounted ThermoPro TP8 thermal-imaging camera (384 × 288 pixels) with a 70 mm telephoto lens at night. They found that waterbirds were active at night and, notably, that the waterbirds used different habitats at different times during a 24-hour period. These results could have huge implications for conservation: where surveys only focus on daytime activity, vital data can be missed. In this case, such results could lead to poor wildlife management decisions and may have a serious impact on waterbird populations. The title of Austin et al. (2016)'s paper, 'If waterbirds are nocturnal are we conserving the right habitats?', raises an important question, one that might also be transferred to a whole host of other species.

6.2.3 Birds in flight

Over the years several researchers have used thermal imaging to directly detect birds in flight in order to study bird migration (Anderson et al. 2021, Fortin et al. 1999, Gauthreaux and Livingston 2006, Liechti et al. 1995, Liechti et al. 2003, Zehnder et al. 2001). Most recently, Anderson et al. (2021) used the technique to study the flight directions of migrating passerines crossing the southern coast of Lake Erie, Ohio, USA. They used a ground-based tripod-mounted FLIR SR-19 thermal-imaging camera, which they oriented vertically to detect birds flying overhead overnight. This device had a resolution of 320 × 240 pixels and a 36° field of view, which they estimated would give them a 'survey volume' (maximum detection distance) up to 600 m above ground level based on prior

Case study: 'Thermal Birding' by West Midlands Bird Ringing Group (WMBRG)

Bird ringing is carried out to gain information on bird movements, survival and productivity. Such data is vital to inform conservation efforts. West Midlands Bird Ringing Group (WMBRG) work with a range of bird species using thermal imaging, including:

- Nightjar *Caprimulgus europaeus*
- Lapwing *Vanellus vanellus*
- Tawny Owl *Strix aluco*
- Skylark *Alauda arvensis*
- Woodcock *Scolopax rusticola*

In 2021, the WMBRG received The Marsh Award for their cutting-edge thermal-imaging work monitoring farmland birds, putting them at the forefront of the use of this technology for this species group.

How has thermal imaging changed their approach?

Traditionally, prior to using thermal imaging, WMBRG used a method known as 'dazzling'. This involves walking around fields on foot at night with a torch searching for birds, which can then be netted by hand. This process can be considerably time-consuming and inefficient. Now, however, they use handheld thermal-imaging scopes (including models from the Pulsar Helion XQ range) to help them find birds by scanning their survey areas and then honing in on their target animal(s). This has transformed the way they do their work and the results they can achieve.

What benefits has it given them?

Introducing thermal imaging to their bird work has brought many benefits, including:

- reduced disturbance
- increased accuracy
- better efficiency

More specifically, it allows them to take a much more focused approach to finding birds in the field, meaning that they save time and effort in comparison with the more traditional approach. It may also mean that they have more time to focus on the actual data they are collecting rather than the process of finding the birds to get it.

For more information on Thermal Birding, visit West Midlands Bird Ringing Group (WMBRG)'s website at: https://www.westmidlandsringinggroup.co.uk/thermal-birding

results. They connected their thermal device up to an external laptop in order to view and manually noted the times of each passing bird. Having manually referenced the direction of the camera using a compass, they were later able to manually review each flight event to note the directions of flight for each passing bird.

Focusing on another topic, Syposz et al. (2021) used thermal imaging to detect avoidance behaviours of seabirds in relation to different artificial light stimuli (see more on this in Section 5.3).

6.3 Bats

With over 1,400 species worldwide (Simmons and Cirranello 2021), the Chiroptera are a diverse order of mammals occupying a wide range of ecological niches. Bats are vital components of healthy ecosystems and are often considered to be important indicator species (Zukal, Pikula and Bandouchova 2015; Conno et al. 2018; Tuneu-Corral et al. 2020). However, with the power of flight and primarily nocturnal activity in a large proportion of species, they are an inherently challenging group to study (Kunz and Parsons 2009). Traditionally, overcoming these challenges has involved a mixture of different techniques to detect and count these fascinating creatures including: manual roost counts, bioacoustics, infrared reflectance, night vision, capture-mark-recapture and, more recently, thermal imaging (Kunz and Parsons 2009).

Early studies pioneered the use of thermal imaging for bats in the 1990s (Kirkwood and Cartwright 1991; Kirkwood and Cartwright 1993; Sabol 1995; Lancaster 1997). After initial trials using thermal imaging for bats in a laboratory setting (Kirkwood and Cartwright 1991) the same team of researchers in the USA took the technique out into the field (Kirkwood and Cartwright 1993). They tested the use of an Inframetrics 525 infrared imaging radiometer and a night vision system (Star Tron MK 426) to record Big Brown Bats *Eptesicus fuscus* inside their maternity roost within a church building and during emergence from it. While they were able to detect bats with it, their thermal-imaging system was bulky and impractical, requiring a heavy and somewhat cumbersome combination of the device, a control unit, a TV monitor, batteries, a VCR recorder and a trolley to move it around. Once recorded onto VHS videotape, the videos and still images could be reviewed and analysed manually.

Thankfully, since the 1990s, technology has come a long way. Advances in the portability, powering and general usability of thermal devices mean that their applicability for bat applications had improved dramatically over the decades. However, these benefits have brought along with them the challenge of the growing mountains of data generated. This is a particularly acute issue when attempting counts of larger colonies that can aggregate in their thousands (Frank et al. 2003; Betke et al. 2008; Hvristov et al. 2010; Matzner 2015). Even with smaller numbers of bats, manual analysis of this kind of data can be tedious and time-consuming. Thankfully, researchers have approached this application and worked to address this challenging issue with the use of automated analysis procedures (Frank et al. 2003; Betke et al. 2008; Hvristov et al. 2010; Cullinan 2015; Matzner 2015; Corcoran, A. J. et al. 2021). Until recently, automated procedures have been largely inaccessible to bat professionals outside of academia, requiring specialist software and programming skills. This has recently changed thanks to the work of Corcoran, Schirmacher, Black and Hedrick (2021) who have created an open-source tool called ThruTracker which is now freely available online (you can read more on this in Chapter 3).

In many parts of the world, bats are at risk due to displacement, injury or death associated with wind energy production (Newson et al. 2017; Gaultier et al. 2020). Researchers focusing on this issue have used thermal imaging to detect and track bats around wind turbines (for more on this see Chapter 5) allowing them to document events that would be otherwise impossible.

Yang et al. (2013) used a combined technology approach to studying bat flight behaviour in relation to their surrounding habitat. Their equipment included two FLIR ThermoVision SC8000 thermal-imaging cameras and an Echidna Validation Instrument (EVI) ground-based scanning LiDAR. The SC8000 model of thermal-imaging camera used in this study are high-end cooled devices, providing high-resolution (1024 × 1024 pixels) thermal imagery at high frame rates (over 130 Hz).

Using such thermal-imaging devices alongside LiDAR (Light Detection and Ranging) they were able to reconstruct the 3D flight paths of Big Brown Bats from thermal-imaging video data contextualised within 3D representations of the surrounding environment (roost building and forest vegetation) from LiDAR data. While this approach was cutting edge for its time, it is not unreasonable to hypothesise that one day this approach could be widely used for both research and survey applications.

Whereas most thermal-imaging studies on bats have employed stationary systems to detect bats in flight, McCarthy et al. (2021) used a drone-based approach to count stationary Grey-headed Flying Foxes *Pteropus poliocephalus* in their tree roosts. In this Australia-based study, researchers used a Zenmuse XT thermal camera (resolution = 640 × 520 pixels) carried by a DJI Inspire 1 Version 2.0 drone. They incorporated automation into their procedures for both data collection and analysis. Data collection was automated using the flight planning application Pix4Dcapture 4.9.0, which allowed take-off, pre-planned flight cruising and landing without manual intervention. The resulting orthomosaics could then be analysed manually and/or using a range of different semi-automatic procedures. Their results suggest that the use of such methods can yield accurate count data for roosting flying foxes and that this can be a valuable tool to monitor their numbers.

Bats may be the group for which using thermal imaging bring some of the most substantial benefits, but there are some key requirements that enable successful implementation (Fawcett Williams 2021). For example, it is particularly critical to select appropriate thermal-imaging equipment for bat applications. Berthinussen and Altringham (2015) tested a number of potential methods of monitoring bats around infrastructure projects, including a thermal-imaging device that was unsuitable for such purposes. Accordingly, this equipment provided them with very poor results. In recent years, methods have since been developed to detect and count bats using thermal imaging. This has allowed appropriate use of the technology that yields better accuracy and efficiency than traditional methods have previously allowed. Guidelines have also become available for its use in professional bat survey work (Fawcett Williams 2021). You can find more details and key information from literature including the use of thermal imaging for bats between 1968 and 2021 in Table 4 in the Appendix.

6.4 Marine mammals

The presence of water can be a major issue for thermal imaging. First, water is highly reflective of infrared waves, meaning we cannot see through the surface of the water. Second, the presence of water on the skin, fur, or feathers of an animal affects the emissivity value of the surface we may wish to detect or measure.

Aside from the ocean water itself, another major challenge in imaging marine mammals is the environmental conditions above the water. These conditions can greatly impact the practicalities of deployment and the thermal considerations for the imagery itself, which can have a major effect on detectability.

6.4.1 Cetaceans

As we might expect, spending the majority of their lives fully submerged in water, cetaceans are probably the most challenging group of marine mammals to detect with thermal-imaging technology. Yet, this is not an impossible feat. Thermal imaging may not be able to 'see' through the water to detect submerged bodies of cetaceans, but it has instead been able to detect them as they breach the surface of the water. Some researchers have successfully used thermal imaging to study this tricky group of marine mammals by focusing on parts of their bodies that emerge from the water or on their signs (such as blow and flukeprints) as they come up for air.

When Barbieri et al. (2010) studied surface temperature in Bottlenose Dolphins *Tursiops truncatus* in Florida, they were able to distinguish the dorsal fins of these animals in the wild. As this study was conducted in parallel to photo-identification studies and with the goal of recording the temperature measurement of the dorsal fins, it is not clear how often or easily this dorsal fin distinction could be made. The authors went to great lengths in testing the reliability of the FLIR Agema 570 IR thermal-imaging camera – using simulations and trials on captive individuals – prior to their field sampling. However, from their methods, it appears that they may have been using the apparent temperature rather than a corrected or estimated surface temperature, which was compared to water temperatures measured using another device. The temperature differences they report (corrected or otherwise) between the dorsal fin and the water were between 0.9 °C and 4 °C. However, it would be difficult or even impossible to detect temperatures of less than a degree with the equipment they used and would likely be difficult or even impossible to use for detection, depending on the surrounding environmental conditions and other key factors. From these results, it remains unclear as to whether dorsal fins might be a reliable or useful target when using thermal imaging for detection purposes.

On the other hand, thermal imaging has been demonstrated to be useful for detecting signs of cetaceans as they breach the surface. Thermal patterns are created by the blow of whales as they exhale and by their movement in the water. These patterns indicating the presence of cetaceans have successfully been detected using thermal imaging for multiple species.

When Cuyler (1992) set out to test thermal imaging as a new technique to look for Minke Whales *Balaenoptera acutorostrata*, they were also able to distinguish several other cetaceans – Fin Whale *Balaenoptera physalus*, Blue Whale *Balaenoptera musculus*, Humpback Whale *Megaptera novaeangliae* and Sperm Whale *Physeter macrocephalus* – as they breached the surface. Using an Agema 880 thermal-imaging camera on board a 30 m boat, they found only small differences between temperature readings of the small areas of emerging body surfaces of the whales and those of the surrounding water. This, they concluded, provided little scope for successful detection. Fortunately, however, they found that the blows of the whales were much more distinct and provided a much better target feature for detection purposes. These could be seen at long range, with the large-scale blows of Blue Whales even being picked up at the horizon (estimated by the authors to be detectable by the camera ~1 km from the boat). However, the use of the equipment was not without its challenges in this study, and the authors reported that successful detection was reliant on favourable weather conditions.

In another on-board application, Baldacci et al. (2005) conducted a feasibility study to assess the potential of thermal-imaging devices for marine mammal detection. They used a tripod-mounted Sagem MATIS binocular device designed for military applications (Sagem Defence Securite, France) from the deck of a NATO research vessel. In favourable conditions, they were able to detect blows of larger cetaceans, emerging fins and flukes, and disturbances in the water caused by these animals' movements. However, they emphasise how severely the weather affected the operation of this equipment, rendering it unusable in a range of sea and weather conditions.

Also on board a ship, but even more recently, Burkhardt et al. (2012) used a very different device to detect whales than that used by Cuyler's team back in the 1990s or by Baldacci et al. in the early 2000s. Mounted on a stabilising gimbal on the crow's nest of a large research vessel, the FIRST-Navy cooled thermal-imaging sensor (Rheinmetall Defense Electronics, Bremen, Germany) used by Burkhardt's team allowed them to detect, locate, and track cetaceans from thermal signatures of their blows at up to 3 km away. This device, with its spinning detector, provided an incredible 360° field of view. From the manual analysis of the data generated, the team was able to reveal signs of five different species – Minke Whales, Humpback Whales, Sperm Whales, Fin Whales and Bowhead Whales – by visually inspecting thermal images after sampling. Later, this same team further developed this method to include the use of automation to streamline their analysis procedure (Zitterbart 2013). The overall procedure allowed them to efficiently obtain and process long-range (up to 5 km), around-the-ship and around-the-clock data that could not be obtained using traditional Marine Mammal Observation (MMO) techniques.

In their study of migrating Grey Whales *Eschrichtius robustus* off the west coast of the USA, Perryman et al. (1999) used a shore-based thermal-imaging set-up to detect and count individuals passing through their study area during migration events. They used tripod-mounted AN/KAS-1A devices facing out to sea as the basis for their thermal-imaging systems and recorded thermal video data to VHS tapes. Using a manual video analysis procedure, they found that the blow of the whales was clearly distinguishable in their thermal-imaging videos, enabling long-range detection (over 4 km) of individual respiration events. This allowed the researchers to extract a range of useful information from the data they collected over a three year period, including migration rates. The use of thermal imaging allowed them to study whale migration rates both during the day *and* at night. This is something that could not be achieved with the visual shore-based techniques typically used for this purpose, meaning that previous estimations were probably inaccurate.

Another shore-based approach was taken by Graber (2011) and Graber et al. (2011) in their work on the detection of Orca *Orcinus orca* off the west coast of the USA. They mounted a FLIR ThermoVision A40M thermal-imaging camera (alongside two other visible spectrum cameras) on the railings of a lighthouse overlooking the Haro Strait where orca are commonly seen (Graber 2011). From their thermal-imaging data, they achieved successful detection of the emerging bodies, dorsal fins, and blows of orca at a range of 43–162 m. Detection of the signatures of body and fin were limited to shorter ranges, whereas blow was apparent from further away. The latter was therefore primarily relied on to detect and identify orca at distances over 100 m in this study.

The most recent – and perhaps most interesting – sign of cetaceans described so far using this approach, has been that of thermal flukeprints produced by Narwhal *Monodon monoceros* (Florko 2021). These flukeprints are patterns created by the movement of surface water as the animals breach and descend during a breathing event. One of the useful things about using flukeprints for detection is that they have a higher latency than the blow or the surface breach events. As the thermal signature lasts much longer, the window of opportunity for detection is longer, which may provide a higher chance of success. In their Arctic based study, Florko (2021) used a combined-technology aerial approach to detecting their target species, using a FLIR T1020sc thermal-imaging camera alongside a Nikon D810 digital single-lens reflex (DSLR) camera, looking down from the underside of a small aeroplane. This combination allowed them a bird's-eye view, with the benefits of using both visible and infrared wavelengths. They were able to use the thermal-imaging data to rapidly detect signs of the animals, and then refer to the photographs from their DSLR to confirm that Narwhals were present and to count the number of individuals, which could be completed faster and more accurately than by using photographs alone.

Naturally, the major limitation of using thermal imaging for this group of marine mammals is its inherent availability bias; it can only detect the animals or their signs during the brief windows of time when individuals are at – or close to – the surface of the water. These fleeting opportunities only occur during or around respiration events, which vary in the duration of the event itself and the duration of the interval between events. Respiration events may vary depending on the species, dive depth, and individual behaviours. Despite this limitation, more recent studies suggest that thermal imaging can provide useful data that can be collected and processed in an efficient and reliable manner.

6.4.2 Pinnipeds

Compared to the cetaceans, pinnipeds offer a relatively easy opportunity for detection when they are hauled out on land or ice (for example, see Figure 19). Seizing this chance to pick up the thermal signatures of these animals while they are out of the water, researchers have used thermal-imaging techniques for seals and walruses. Of course, this opportunity only allows for detection of the animals hauled out at the time of sampling, which may be biased towards particular sex and/or age classes (Kingsley 1990).

The innovators in this field used aircraft-mounted thermal-imaging devices to find walruses on ice (Barber et al. 1989; Barber et al. 1991). Early attempts to use thermal imaging for pinnipeds gave mixed results, but researchers saw considerable potential in this application and pursued its development despite the initial challenges they encountered. Since then, in more recent studies, with the advantages of more advanced hardware and software, early adopters that went on to further develop this technique have demonstrated it to be a useful tool (Burn et al. 2006; Udevitz 2008; Burn et al. 2009).

While pinnipeds may usually be readily available for detection when hauled out, it is important to note that the event of 'hauling out' in itself may affect detectability. In an excellent review paper, McCafferty (2007) illustrated how water on the surface of mammal fur can affect detection results. When pinnipeds have recently hauled out, water can change the way they look in a thermal image because it affects the emissivity (which, in turn, affects apparent temperature), evaporative cooling, and thermal conductivity of their body surfaces (McCafferty 2007).

In a largely simulation-based study, Conn (2014) provided a real-life example of data on seals hauled out on ice in the Bering Sea. Using aerial methods, the researchers used three FLIR SC645 thermal-imaging cameras mounted on a plane, which was followed by another plane fitted with visible-light cameras. They used an automated detection system to find seals in their thermal data and compared these detections to manual detections identified from their aerial photographs. From their thermal imagery, they discovered that they could automatically detect 94% of the seals found manually in photographs. The latter task took ~120 hours to complete, whereas the former was likely conducted in minutes (though this is not detailed specifically in the paper).

In order to count Grey Seals hauled out on land, Seymour et al. (2017) created an efficient workflow process to automate both their data collection and their data analysis. Focusing on breeding colonies of seals in North America, they collected their data using a senseFly Thermomapper drone-mounted thermal-imaging camera (senseFly LLC, Raleigh, NC, USA). They automated their analysis procedure with a model built using ArcGIS software (Esri (UK) Ltd. Aylesbury, UK). This allowed them to count and classify seals based on key features from the thermal images they collected, using mechanisms including filtering and edge detection. Classification aimed to assign detected seals into one of two categories: adult or young of the year. When they compared their automated counts to manual counts using the same data, their automation counts were slightly lower than those made by humans (with automation achieving ~91%, ~95%, ~96% and ~98% of that achieved by human counts at each of their study sites respectively). It appears that automation very rarely missed adult seals, but errors were encountered for pups. Although not documented specifically in their paper, it seems highly likely – judging from the procedures involved – that the difference in the time and associated costs between these two approaches could be vast.

Seymour et al. (2017) discuss the importance of timing to maintain accuracy; timing can be optimised to reduce the likelihood of both false positives and false negatives. First, time of day was important in their study. It was necessary to select times when there was a good thermal contrast between seals and their surrounding habitat, otherwise seals could be 'lost' in their surroundings, resulting in false negatives. False positives were also an issue at certain times of day, caused by non-target landscape features that had been warmed (or were reflecting infrared waves) and could be wrongly identified as a seal. To avoid such scenarios, the team identified early dawn as an optimal window for sampling, while taking caution that other considerations may be necessary: whether light will be sufficient for the use of an RGB camera, and the behaviour of the seals themselves. In the latter case, it is important to consider site-specific haul-out behaviours to ensure that seal numbers are not underestimated.

In a study comparing aerial survey techniques for counting Grey Seals, Johnston et al. (2017) used a senseFly Thermomap 1.2MP drone-mounted thermal-imaging camera (senseFly LLC, Raleigh, NC, USA). Unfortunately, they were not able to use this camera consistently for logistical rather than technical reasons, so they do not share the details of their results in their paper. However, they do speculate on the potential the technology has for this application.

Focusing on New Zealand Fur Seals *Arctocephalus forsteri*, Gooday et al. (2018) tested the effectiveness of ground-based and drone-mounted thermal-imaging cameras to help them count individuals in different and arguably more challenging habitats than we have discussed so far for this group of marine mammals. Not only do these animals haul out onto land to give birth and feed their pups, but they also take shelter under tree cover of varying density and complexity. This presents a challenging environment for researchers to monitor seal numbers, and new methods were required in order to do this in a safe and non-invasive way. To test detecting and counting from the air, the researchers used a drone-mounted T320 thermal-imaging camera alongside a visible-light camera. To assess another perspective, they also used a tripod-mounted ground-based system with an Optris PL450 thermal camera. They were able to view images from these cameras in real-time on

Figure 19 Hauled-out seals at Donna Nook, Lincolnshire, UK. Timing is important for most applications we discuss in this book, but it appears to be crucial for the successful use of thermal imaging of hauled-out pinnipeds.

a monitor screen during sampling, and to process and/or analyse any of their footage after sampling was complete. For the aerial surveys, detection rates were poor in comparison to ground-based counts made by human observers on foot. Under the most challenging conditions of 95% canopy cover, they were unable to detect any of the seals present in the manual ground counts. Under 50% canopy cover, they were able to detect 17–68% of the seals counted in the manual ground counts. The results from their ground-based system were much more promising: thermal imaging outperformed digital cameras using visible light in 11 out of 18 scenarios and was equally effective in 4 out of the 18 scenarios. Sampling at different times of day gave very different results for both aerial and ground-based thermal-imaging systems, reinforcing the importance of timing for applications such as these.

In an Antarctic environment, Hyun et al. (2020) were able to detect and count Southern Elephant Seals *Mirounga leonina* when they were hauled out onto the gravel shore of King George Island. Using images collected with a drone-mounted FLIR Vue Pro R thermal-imaging sensor, they were also able to extract quantitative features by analysing their data using ArcGIS software. Based on their findings, they suggest that it would likely also be possible to distinguish between age and sex classes using this kind of data.

Recent studies have shown that it is possible to use thermal imaging for the detection and enumeration of pinnipeds hauled out on land or ice. While this provides a great window of opportunity, timing is crucial to obtaining accurate results on this group of marine mammals under such circumstances. When using such methods, wildlife professionals should account for environmental conditions as well as hauling out behaviours that can affect detectability. Testing should always be carried out prior to sampling in new areas or colonies to verify site-specific optimal conditions.

Case study: Counting Atlantic Grey Seal pups (Farne Islands, UK)

In autumn, female Atlantic Grey Seals haul out onto the shores of the Farne Islands in the UK to give birth to their young. Rangers at the National Trust are tasked with counting the number of these pups born on the islands each year (National Trust 2021). With a record 2,823 pups born in 2019, this is no mean feat.

How do they usually count them?

Rangers have traditionally counted seal pups using a very low-tech procedure. During the breeding season, they make regular visits to the breeding seals on foot, counting each new arrival. During the process, they mark each newborn seal with vegetable-based dye. This is not only labour intensive, but it also involves human interaction with the seals which has potential implications in terms of disturbance. In recent years, conditions for manual counts have become more and more difficult due to weather events and the changing distributions and densities of seals on the islands.

How are they using thermal imaging?

In collaboration with researchers at Newcastle University, Oxford University, the Sea Mammal Research Unit (SMRU), and TerraDrone, the National Trust researchers have been trialling the use of thermal-imaging cameras alongside RGB cameras deployed using drones to collect their seal data. Using a DJI M300 RTK drone with a Zenmuse H20T thermal-imaging camera and on-board RGB camera, they were able to view thermal and visual data simultaneously in real-time, which can be extremely valuable in the field (Heliguy 2021). Following data collection, they have also used a machine-learning approach to automate their data analysis procedure.

What are the benefits?

In comparison to the usual manual methods, using this combination of technology brings a suite of new benefits to this seal counting application, including:

- higher efficiency
- greater accuracy
- cost-effectiveness
- lower invasiveness

By automating both their data collection and data analysis, the researchers have made this a very efficient workflow process. Not only has automation made the process much faster and more cost-effective, but it has improved accuracy by picking up more seals than manual counts. More importantly, perhaps, it also allows data to be gathered in a much less invasive way, which may have huge implications for the welfare and survival of these animals at such a critical point in their lives.

For more information on the use of thermal imaging to count seal pups on the Farne Islands, visit the National Trust's website: https://www.nationaltrust.org.uk/press-release/thermal-imagery-used-for-first-time-to-help-count-seal-pups-on-the-national-trusts-farne-islands-

For more information on the technical details of drones used in this work, see the Heliguy website: https://www.heliguy.com/blogs/posts/m300-rtk-drone-seal-pup-count

6.4.3 Polar Bears

Threatened by habitat loss as a result of climate change, oil and gas operations, and other human activities, populations of Polar Bear *Ursus maritimus* are declining (IUCN 2015). The first marine mammal to be studied using thermal imaging back in the early 1970s (Brooks 1972), Polar Bears have been the focus of a small number of subsequent studies implementing this technology to detect these animals and/or their dens in the wild (Amstrup et al. 2004; York et al. 2004; Robinson 2014; Smith, 2020). In the very early stages of the testing of this technology, Brooks (1972) failed to detect Polar Bears with the primitive thermal-imaging device they used on board a plane. However, on further inspection of their aerial imagery during manual analysis, they found that they could see thermal trails left by the Polar Bears in the snow.

Following this early work, seemingly little was done on the subject, until over three decades later when two publications appeared on the topic in the same year (Amstrup et al. 2004; York et al. 2004). In the spring of 2004, a 'Biologist's Toolbox' article was published detailing the use of a helicopter-mounted FLIR Safire II, AN/AAQ-22 thermal-imaging device to test its effectiveness for the detection of what they refer to as heat rising from Polar Bear dens (Amstrup et al. 2004). Polar bear dens are excavated in the snow by pregnant females, and this is where they give birth to and raise their cubs. The protection of these dens could be vital to the survival of this species, and so being able to detect them is critical to their protection. Amstrup et al. (2004) reported that they achieved almost 90% accuracy when looking for dens using this method, but that this relied on the work being conducted under optimal conditions. In the autumn of 2004, a comprehensive operations manual on the subject was published (York et al. 2004). This 58-page document reviews in detail the testing procedures that had been carried out by the authors – seemingly a more detailed account of the content in Amstrup et al. (2004) – and provides in-depth guidance on the entire process of using aerial thermal-imaging techniques to detect Polar Bear dens, including the all-important conditions necessary for its appropriate use.

To investigate the efficacy of thermal imaging to detect Polar Bear dens, Robinson et al. (2014) created mockups of their target feature and tested their detectability using a handheld FLIR ThermaCAM under different conditions. While this study did not include any data on real Polar Bear dens, it did emphasise some of the major issues of using this approach. They found that weather conditions affected performance, with solar radiation and wind speed being the most notable influences on success or failure. From their findings, they concluded that surveys of this nature should be carried out overnight under low wind speeds to give the best chance of detecting dens. More importantly, they found that the depth of dens was a major factor, with dens more than 90 cm below the surface going undetected. In practice, this means a substantial proportion of dens are likely to go unnoticed when this survey method is used.

Reading between the lines, it would seem that since the work of Amstrup (2004) and York (2004) formally introduced the method of thermal imaging to find Polar Bear dens, it has been used by consultants and wildlife managers to survey and monitor this species. However, the results of such work have most likely been confined to the pages of confidential reports until Smith et al. (2020) examined the results of 13 years of data collected in relation to oil industry exploration activities in Alaska. They evaluated the work of industry professionals collecting aerial thermal-imaging data to help inform operations to minimise disturbance to Polar Bears. Using ground-based data of known Polar Bear dens as a comparison, they found that aerial surveys were achieving poor rates of detection (45%). On closer inspection of the methods used to achieve this, they found that data collection did not follow the necessary protocols outlined in the research and guidance (Amstrup et al. 2004; York et al. 2004) on the subject, which meant their results were largely invalid. Specifically, operators had been using the technology under unsuitable weather conditions, which has been clearly demonstrated to have a huge bearing on detection accuracy. Regardless of these conditions, however, they also note the limitation of snow depth on den detection, in accordance with the findings of Robinson et al. (2014). Due to the poor outcomes from the vast amounts of data they reviewed, Smith et al. (2020) called for alternative methods to gain more reliable data, and stated that protocols should be strictly adhered to when conducting surveys.

On reflection, from the literature we have discussed so far, it appears that thermal-imaging technology has been used somewhat inappropriately for Polar Bears. As efforts have been met with mixed results in detecting signs (trails and dens) of Polar Bears, it would seem that this approach is realistically neither appropriate nor reliable. To detect Polar Bears effectively, there is a need to refocus thermal-imaging efforts or to find alternative methods altogether. The former case requires testing of newer devices to detect these animals directly. This, of course, is not without its challenges: Polar Bears are heavily insulated by thick and specially adapted fur which can act as an invisibility cloak to thermal-imaging cameras (Cui 2018). However, since the days of Brooks (1972), these technologies have come a long way; given the substantial advances, direct detection is worthy of further research to evaluate its usefulness.

7. Future Possibilities

While thermal-imaging technology is now being used more than ever before for wildlife applications, it is yet to be fully utilised in the mainstream. There are so many possibilities for this technology to be employed for a wide range of species and applications in the future. In this final chapter, we will consider some of the potential developments that we might expect the future to hold for this exciting field.

7.1 Further development of thermal-imaging technologies

In recent decades we've seen a dramatic increase in the use of thermal-imaging technology for wildlife applications. This has been facilitated by the progress made in the global development of thermal-imaging technologies. Improvements in the availability, affordability, portability and user-friendliness of thermal-imaging devices have been, and will continue to be, instrumental to the applicability of thermal-imaging devices for wildlife work in the field.

As outlined in Chapter 4, thermal detector resolution is a key specification that defines the suitability of a device for a specific wildlife application. This parameter determines the detection distances we can achieve and the quality of images we can acquire, both of which can have huge implications for our data. Increases in resolution seen in recent decades have naturally led to greater use of thermal imaging for global applications where higher quality images or longer-range detection is required. Most people that have used thermal imaging so far still eagerly await the ever-increasing number of pixels promised by future projections for this technology. Further improvements to resolution capabilities will naturally open up greater possibilities for a more diverse range of wildlife species applications in future.

7.2 Automation

The area of development that has the biggest potential to expand the use of thermal-imaging technology for wildlife applications is automation. Currently, thermal-imaging data collection and subsequent analysis tasks can require an investment of human labour and associated costs. Automating these processes is key to the scalability and applicability of this technology for a wider range of wildlife applications.

7.2.1 Automation of data collection

Automating data collection could dramatically increase efficiency and reduce the costs of fieldwork. This could be done in a number of ways, although some are more practical than others. Further development of drone-based applications through the improvement of both technology and methodology could vastly expand this mode of automation of data collection.

Theoretically, data collection could be automated by using remotely operated or static thermal-imaging devices. These might take the form of camera-trap-style devices, security camera style systems or custom systems created with commercially available equipment. Depending on the type and specification of the device, this could allow automatic data collection over a range of recording timescales: from one-off surveys through to long-term monitoring programmes. Unfortunately, in practice, the current financial value of most thermal-imaging equipment means that very few, if any, wildlife professionals would be willing or able to leave it *in situ* to collect data unattended. Likewise, many thermal-imaging devices are strictly controlled, which can severely limit or prevent this technology from being deployed in such a way. Therefore, any future system of this type would require the ability to easily secure and protect valuable technology, or alternatively involve the development of a low-cost system that could sensibly be left unattended in the field.

Security-style camera systems are well developed and established, but these are not often available for purchase by the general public. Such systems have many desirable attributes for use in the long-term monitoring of wildlife species, including long-range detection capabilities, weatherproofing and the ability to operate remotely from a central control system. If such technology could be made available to wildlife professionals, this would open up a valuable opportunity to test this type of equipment and ground truth what appears on paper to have great potential implications for the detection and monitoring of wildlife.

The creation of commercially available thermal camera traps would be a particularly exciting development to automate data collection. This has been done experimentally for research applications, but there is no sign to suggest that such devices will make an appearance on the market anytime soon.

As a practical alternative, we can also use existing technologies to create custom systems to automate the collection of thermal-imaging data for wildlife applications. All of the components for such systems are readily available, but the expertise to connect them up into a working system is often difficult to find among wildlife professionals. Often the best technical solutions in situations like this have come from working collaboratively with professionals from other disciplines where such expertise is present in abundance.

7.2.2 Automation of data analysis

We are collecting more and more data year on year, and much of this data is useless unless it is analysed in some way. Until relatively recently, this data has been analysed manually within specialist software packages. But as the volumes of data continue to increase, large-scale fully manual data analysis will become unrealistic at some point in time. With the exception of specialist research applications for researchers with access to custom-built programmes and the necessary skills to use them, the automation of data analysis has been largely inaccessible to the majority of wildlife professionals using thermal-imaging technologies.

In the future, wildlife professionals using this technology will need access to user-friendly automation software that allows them to process their data efficiently and extract meaningful results in a timely manner. It is also critical that wildlife professionals using such software understand the basic mechanisms of the automation processes they use, and both the capabilities and limitations they entail.

7.3 Data storage and handling

In recent years we have seen much improvement in the cost and capacity of data storage. Slowly but surely, this has allowed us to store more data in more efficient and cost-effective ways. However, simply increasing the size and affordability of storage available will not solve all the problems associated with wildlife data needs.

As the volume of wildlife data we collect will likely continue on a steep upward trajectory in future, we desperately need better systems, processes and skills to effectively manage the data we collect.

7.4 Sustainability

The increased use of thermal imaging and other technologies for wildlife applications can provide huge benefits to both wildlife and humans. Unfortunately, at the time of writing, there are also significant environmental costs involved. These costs are associated with the acquisition of raw materials, production of devices and their component parts, transportation, long-term running costs, recycling and/or disposal.

When we are working in wildlife applications that are vital to the preservation of species, particularly those whose numbers are low or are even on the verge of extinction, we can become laser-focused on using whatever resources we can to do the work required to help them. While this is understandable, it is important to be mindful of the global impacts of our use of technology and work out how we can do better. While continuing to innovate in this field, we also need to develop more sustainable ways of using technologies like thermal imaging in our work.

Thermal-imaging devices in particular contain some rare and valuable materials that are not only hard to come by but, like any of our earth's resources, are inherently finite. As a technology that is developing and growing, sustainability should be a key consideration for the future of the thermal-imaging industry. This could include a transition from its current reliance on the extraction of raw materials, towards a more circular approach to the supply of materials for device production. Likewise, improvements in the energy efficiency of devices and systems are also needed to improve their sustainability.

By adopting more circular economic models, the thermal-imaging industry could dramatically improve the sustainability of material acquisition and device production. This may subsequently reduce the need for mining of raw materials such as germanium by reusing and/or recycling existing components from devices that have reached the end of their service.

Appendix

Table 1 Literature including the use of thermal imaging for wildlife applications between 1968 and 2021 (inclusive).

Year	Lead author	Title	Target species group	Device type	Make	Model	Deployment type
1968	Croon	Infrared Scanning Techniques For Big Game	Deer	Line Scanner	Stoll-Hardy	Radiometer	Aerial (Plane)
1972	Brooks	Infra-Red Scanning for Polar Bear	Polar Bear	Scanner	Daedalus	Detector (In:Sb)/ Trimetal detector (Hg:Sb:Te)	Aerial (Plane)
1972	Graves	Censusing white-tailed deer by airborne thermal infrared imagery	Deer	Line Scanner	Not specified	Detector (In:Sb)/ MMercury Cadmium Telleride (MCT)	Aerial (Plane)
1972	Parker	An Experiment in Deer Detection by Thermal Scanning	Deer	Scanner	Bendix	LN-2-LW	Aerial (Plane)
1973	Cena	Thermographic measurements of the surface temperatures of animals	Zebra, Elephant and Giraffe	Camera	Agavision System	680	Vehicle-mounted
1977	Wride	Thermal Imagery for Census of Ungulates	Moose, Deer, Elk, Bison	Line Scanner	Daedalus	Not specified	Aerial (Plane)
1982	Best	Aerial thermal infrared census of Canada geese in South Dakota	Geese (Canada)	Line Scanner	Daedalus	Not specified	Aerial (Plane)
1989	Barber	Thermal remote sensing for walrus population assessment in the Canadian arctic	Walrus	Not specified	FLIR	Not specified	Aerial (Plane)
1990	Kingsley	Infrared sensing of the under-snow lairs of the ringed seal	Ringed Seals	Infrared Sensor	FLIR	100A	Aerial (Helicopter)
1991	Kirkwood	Behavioral observations in thermal imaging of the big brown bat, *Eptesicus fuscus*	Brown Bats	Not specified	Inframetrics	Imaging radiometer	Not specified
1991	Barber	Calibration of aerial thermal infrared imagery for walrus population assessment	Walrus	Not specified	FLIR	Not specified	Aerial (Plane)

Year	Lead author	Title	Target species group	Device type	Make	Model	Deployment type
1992	Cuyler	Thermal infrared radiation from free-living whales	Minke Whales	Not specified	Agema	Thermovision 880	Boat
1992	Klir	An infrared thermographic study of surface temperature in relation to external thermal stress in three species of foxes: The Red Fox *Vulpes vulpes*, Arctic Fox *Alopex lagopus*, and Kit Fox *Vulpes macrotis*	Foxes (Red, Arctic, Kit)	Camera/Scanner	Inframetrics	525	Ground-based – static
1993	Kirkwood	Comparison of two systems for viewing bat behavior in the dark	Bats	infrared imaging radiometer	Inframetrics	525	Ground-based – static
1993	Sidle	Aerial thermal infrared imaging of sandhill cranes on the Platte river, Nebraska	Sandhill Crane	Not specified	Loral	AN/AAD-5	Aerial (Plane)
1993	Wiggers	Use of thermal infrared sensing to survey white-tailed deer populations	Deer	Scanner	FLIR	2000G	Aerial (Plane)
1994	Boonstra	Finding mammals using far-infrared thermal imaging	Red Squirrel, Arctic Ground Squirrel, Snow Hare, Meadow Jumping Mice	Not specified	Agema	Thermovision 210	Ground-based – not specified
1995	Boonstra	Limitations of far infrared thermal imaging in locating birds	Owl, Woodpecker, Flickers, Mallards, Teal, Longspur, Sandpiper	Not specified	Agema	Thermovision 210	Ground-based – static
1995	Garner	Use of modern infrared thermography for wildlife population surveys	Deer, Moose, Turkey	Scanner	FLIR	2000 A/B	Aerial (Plane)
1995	Havens	The use of thermal imagery in the aerial survey of panthers (and other animals) in the Florida Panther National Wildlife Refuge and the Big Cypress National Preserve	Panther, Deer	Camera	InfraCam	Not specified	Aerial (Plane)

Year	Lead author	Title	Target species group	Device type	Make	Model	Deployment type
1995	Liechti	Quantification of nocturnal bird migration by moonwatching: comparison with radar and infrared observations	Birds (spp. unknown)	Not specified	Inframetrics	LORIS IRTV-445L	Ground-based – static
1995	Ono	Unusual thermal defence by a honeybee against mass attack by hornets	Bees	Camera	Avio	TVS-8100	Not specified
1995	Sabol	Technique Using Thermal Infrared-Imaging for Estimating Populations of Gray Bats	Bats	Scanning radiometer	Agema	782	Ground-based – static
1995	Torgerson	Thermal refugia and chinook salmon habitat in Oregon: Applications of airborne thermal videography	Chinook Salmon	Not specified	Agema	Thermovision 800	Aerial (Helicopter)
1996	Benshemesh	Surveying breeding densities of Malleefowl using an airborne thermal scanner	Malleefowl	Two: Scanner/ Unspecified	Daedalus/ Inframetrics	1240/60/445	Aerial (Plane)
1996	Naugle	Use of thermal infrared sensing to estimate density of white-tailed deer	Deer	Not specified	FLIR	2000AB	Aerial (Plane)
1997	Gill	The use of portable thermal imaging for estimating deer population density in forest habitats	Deer	Weapons targeting device	Pilkington Thorn Optronics	LITE direct view thermal imager	Vehicle-mounted
1997	Lancaster	Wing temperature in flying bats measured by infrared thermography	Bats	Not specified	Agema	Thermovision 880	Ground-based – static
1998	McCafferty	The use of IR thermography to measure the radiative temperature and heat loss of a barn owl (*Tyto alba*)	Owls	Camera	AGEMA	AGA 782	Not specified
1998	Havens	Using thermal imagery in the aerial survey of animals	Deer, Panther	Not available	Not available	Not available	Not available
1999	Fortin	Variation in the nocturnal flight behaviour of migratory birds along the northwest coast of the Mediterranean Sea	Birds (spp. unknown)	Not specified	Inframetrics	LORIS 445L	Ground-based – static
1999	Perryman	Diel variation in migration rates of eastern pacific gray whales measured with thermal imaging sensors	Whales (Eastern Pacific Gray)	Not specified	Not specified	AN/KAS-1A	Ground-based – static

Year	Lead author	Title	Target species group	Device type	Make	Model	Deployment type
1999	Haroldson	Evaluation of thermal infrared imaging for detection of white-tailed deer	Deer	Not available	Not available	Not available	Not available
2000	Dymond	Optimizing the airborne thermal detection of possums	Possums	Camera	FLIR	Prism DS/M2000	Aerial (Helicopter)
2000	Belant	Comparison of 3 devices to observe white-tailed deer at night	Deer	Not available	Not available	Not available	Not available
2001	Focardi	Comparative evaluation of thermal infrared imaging and spotlighting to survey wildlife	Deer (Red, Fallow), Boar, Fox (Red), European Rabbit, Brown Hare	Not specified	Galileo	Unit for Fauna Observation (GUFO)	Vehicle-mounted
2001	Zehnder	Nocturnal autumn bird migration at Falsterbo, south Sweden	Birds (spp. unknown)	Camera	Inframetrics	LORIS 445L	Ground-based – static
2002	Arenas	An evaluation of the application of infrared thermal imaging to the tele-diagnosis of sarcoptic mange in the Spanish ibex (*Capra pyrenaica*)	Ibex	Camera	Elbit	Milcam	Ground-based
2002	Dunn	Using thermal infrared sensing to count elk in the south-western United States	Elk	Not specified	Inframetrics	IRTV-445G MK II FLIR	Aerial (Aeroplane)
2002	Grierson	The use of aerial digital imagery for kangaroo monitoring	Kangaroo	Scanner	Redpath Technical Services	Custom device	Aerial (Aeroplane)
2002	Ovadia	Thermal imaging of House Sparrow nestlings: the effect of begging behavior and nestling rank	House Sparrow	Camera	Inframetrics	760	Not specified
2003	Frank	Advanced infrared detection and image processing for automated bat censusing	Bats	Camera	Indigo	Merlin	Ground-based – static
2003	Galligan	Using a thermographic imager to find nests of grassland birds	Sparrows	Not specified	FLIR	ThermaCAM PM575	Ground-based – roaming

Year	Lead author	Title	Target species group	Device type	Make	Model	Deployment type
2003	Haroldson	Evaluation of aerial thermal imaging for detecting white-tailed deer in a deciduous forest environment	Deer	Not specified	Westinghouse Electronic Systems	WF-160DS	Aerial (Aeroplane)
2003	Liechti	Nocturnal bird migration in Mauritania – first records	Birds (spp. unknown)	Not specified	Inframetrics	LORIS IRTV-445L	Ground-based – static
2004	Bernatas	Sightability model for California bighorn sheep in canyonlands using forward-looking infrared (FLIR)	Bighorn sheep	Camera	FLIR	Westinghouse WesCam DS16 FLIR	Aerial (Aeroplane)
2004	Armstrup	Detecting denning Polar Bears with forward looking infra-red (FLIR) imagery	Polar Bear	Not available	Not available	Not available	Not available
2004	Kissell	An assessment of thermal infrared detection rates using white-tailed deer surrogates	Deer	Not specified	Mitsubishi	IR-M700	Aerial (Aeroplane)
2004	Smart	Monitoring woodland deer populations in the UK: an imprecise science	Deer	Weapons targeting device	Pilkington Thorn Optronics	LITE direct view thermal imager	Vehicle-mounted
2005	Ditchkoff	From the Field: capture of white-tailed deer fawns using thermal imaging technology	Deer	Camera	Raytheon	Palm IR 250 Digital	Vehicle-mounted
2005	Drake	Counting a suburban deer population using forward looking infrared radar and road counts	Deer				Vehicle-mounted and Aerial (Helicopter)
2005	Hodnett	Thermal imaging applications in urban deer control	Deer	Camera	Wescam	DS200	Aerial (Helicopter mounted)
2005	Lavers	Application of remote thermal imaging and night vision technology to improve endangered wildlife resource management with minimal animal distress and hazard to humans	Elephant, Ostrich, Camel, Buffalo, Zebra	Camera	Raytheon	PalmIR Pro	Tripod scaled with normal cameras on same target
2005	Potvin	Testing 2 aerial survey techniques on deer in fenced enclosures: visual double-counts and thermal infrared sensing	Deer	Camera	FLIR	2000 A/B sensor	Aerial (Fixed to Helicopter)

Year	Lead author	Title	Target species group	Device type	Make	Model	Deployment type
2005	Willis	Spatial variation of heat flux in Steller sea lions: evidence for consistent avenues of heat exchange along the body trunk	Sea Lion	Camera	FLIR	ThermaCAM PM695	Ground-based – handheld
2006	Allison	Equipment and techniques for nocturnal wildlife studies	Cranes	Scope and Binoculars	American Eagle	Pocketscope, 500 mm C-mount lens	Tripod mount at shoulder height
2006	Burn	Application of Airborne Thermal Imagery to Surveys of Pacific Walrus	Walrus	Scanner	Daedalus	Airborne Multispectral Scanner	Aircraft (Helicopter)
2006	Butler	Limitations of thermal infrared imaging for locating neonatal deer in semiarid shrub communities	Deer	Camera	Raytheon	PalmIR 250 Digital	Tripod fixed on pickup cab
2006	Dunbar	Use of infrared thermography to detect signs of rabies infection in raccoons (*Procyon lotor*)	Raccoons	Camera	FLIR	ThermaCAM E-65	Not specified
2006	Gauthreaux	Monitoring bird migration with a fixed-beam radar and a thermal imaging camera	Birds (spp. unknown)	Camera	Raytheon	TI	Tripod mount pointing vertically up
2006	Locke	Evaluation of Portable Infrared Cameras for Detecting Rio Grande Wild Turkeys	Turkeys	Camera	FLIR	ThermaCAMt B-20	Aerial (Helicopter)
2007	Hemami	Estimating abundance of introduced Chinese muntjac *Muntiacus reevesi* and native roe deer *Capreolus capreolus* using portable thermal imaging equipment	Deer (Muntjac, Roe)	Weapons targeting device	Pilkington Thorn Optronics	LITE direct view thermal imager	
2007	Sumbera	Patterns of surface temperatures in two male-rats (Bathyergidae) with different social system as revealed by IR-thermography	Mole Rats	Camera	Not specified	AGA 570	Not specified
2008	Betke	Thermal imaging reveals significantly smaller Brazilian free-tailed bat colonies than previously estimated	Bats	Camera	Merlin	Mid infrared cameras	Tripod-mounted
2008	Horn	Behavioral responses of bats to operating wind turbines	Bats	Camera	FLIR	Microbolometer TIR cameras	Tripod-mounted

Year	Lead author	Title	Target species group	Device type	Make	Model	Deployment type
2008	Udevitz	Estimation of walrus populations on sea ice with infrared imagery and aerial photography	Walrus	Scanner	Daedalus	Airborne Multispectral Scanner	Aircraft (Helicopter)
2009	Ammerman	Census of the endangered Mexican long-nosed bat *Leptonycteris nivalis* in Texas, USA, using thermal imaging	Bats	Camera	FLIR	S65	Tripod-mounted
2009	Burn	An improved procedure for detection and enumeration of walrus signatures in airborne thermal imagery	Walrus	Scanner	Argon	Thermal infrared (8.5–12.5 mm) scanner	Aircraft (Turbine aeroplane)
2009	Dunbar	Use of infrared thermography to detect thermographic change in mule deer (*Odocoileus hemionus*) experimentally infected with foot-and-mouth disease	Deer	Camera	FLIR	ThermaCAM EX320	Not specified
2009	Kuhn	Infrared thermography of the body surface in the Eurasian otter *Lutra lutra* and the giant otter *Pteronura brasiliensis*	Otter (Giant, Eurasian)	Camera	FLIR	ThermaCAM B20	Not specified
2009	Lavers	Non-destructive high-resolution thermal imaging techniques to evaluate wildlife and delicate biological samples	Rhino, Otter, Dogs, Butterfly, Elephant, Giraffe	Camera	FLIR; Echotherm	ThermaCAM E320; Thermal Wave Imaging	Not specified
2009	Lazarevic	Improving the efficiency and accuracy of nocturnal bird surveys through equipment selection and partial automation	Golden Plover, Gulls, Geese, Harris Hawk, Rook and various	Imager	FLIR	ThermaCAM P60	Not specified
2009	Marini	Response to human presence during nocturnal line transect surveys in fallow deer (*Dama dama*) and wild boar (*Sus scrofa*)	Deer and Boar	Camera	FLIR	ThermaCAM™ PM545	Not specified
2010	Barbieri	Using infrared thermography to assess seasonal trends in dorsal fin surface temperatures of free-swimming bottlenose dolphins (*Tursiops truncatus*) in Sarasota Bay	Dolphins (bottlenose)	Camera	FLIR	Agema 570 IR	Monopod with regular camera on ship

Year	Lead author	Title	Target species group	Device type	Make	Model	Deployment type
2010	Carr	Comparative woodland caribou population surveys in Slate Islands Provincial Park, Ontario	Caribou	Camera	Agema	Thermovision 1000	Aerial (Plane)
2010	Hristov	Seasonal variation in colony size of Brazilian free-tailed bats at Carlsbad Cavern based on thermal imaging	Bats	Camera	FLIR/Indigo Systems	Merlin	Not specified
2010	Conner	Effects of mesopredators on nest survival of shrubnesting songbirds	Birds	Not specified	Raytheon	PALM IR 250	Vehicle-mounted
2010	O'Neal	Waterfowl on weather radar: applying ground-truth to classify and quantify bird movements	Waterfowl	Camera	FLIR	S-60	Tripod Mount
2010	Reichard	Thermal Windows on Brazilian Free-tailed Bats Facilitate Thermoregulation during Prolonged Flight	Bats	Camera	Merlin	Mid TIR camera	Not specified
2010	Weissenböck	Thermal windows on the body surface of African Elephants *Loxodonta africana* studied by infrared thermography	Elephant (African)	Camera	FLIR	ThermaCAM P60	Not specified
2011	Israel	A UAV-based roe deer fawn detection system	Deer	Camera	FLIR	Tau 640	Aerial (Drone)
2011	Kissell	A technique to estimate white-tailed deer density using vertical-looking infrared imagery	Deer	Camera	Mitsubishi	IR-M700	Aerial (Plane)
2011	Millette	AIMS-Thermal – a Thermal and High Resolution Color Camera System Integrated with GIS for Aerial Moose and Deer Census in Northeastern Vermont	Moose and deer	Camera system	AIMS	Not specified	Aerial (Plane)
2011	Mills	Capturing clapper rails using thermal imaging technology	Clipper Rail	Cameras	L-3 Communications	Thermal-Eye 250D; Thermal-Eye X200xp	Handheld on airboat
2011	Storm	Comparison of visual-based helicopter and fixed-wing forwardlooking infrared surveys for counting white-tailed deer *Odocoileus virginianus*	Deer	Camera	PolyTech	Kelvin 350 II infrared sensor	Aerial (Plane)

Year	Lead author	Title	Target species group	Device type	Make	Model	Deployment type
2012	Franke	Aerial ungulate surveys with a combination of infrared and high-resolution natural colour images	Deer (Red, Fallow, Roe), Boar, Fox, Wolf, Badger	Camera	JENOPTIK	Not specified	Aircraft-mounted (Airplane)
2012	Zhou	Contour based HOG deer detection in thermal images for traffic safety	Deer	Camera	Thermal-Eye	TSCss-FF	Fitted on rotating motor. 180 range
2012	Morelle	Game species monitoring using road-based distance sampling in association with thermal imagers: a covariate analysis	Boar, deer, Red Fox	Scope/Camera	FLIR/JENOPTIK	ThermaCAM™ HS–324/ VarioCAM™	Handheld in car
2012	Steen	Automatic Detection of Animals in Mowing Operations Using Thermal Cameras	Chicken, Rabbit	Camera	FLIR	Not specified	Fitted on mowing tractor
2012	Franzetti	Nocturnal line transect sampling of wild boar (*Sus scrofa*) in a Mediterranean forest: long-term comparison with capture-mark-resight population estimates	Boar	Camera	FLIR	ThermaCAM PM545® and ThermaCAM™ B640	Ground-based – roaming
2013	Department for Environment Food and Rural Affairs	Monitoring the humaneness of badger population reduction by controlled shooting	Badger	Not specified	Not specified	Not specified	Ground-based – roaming
2013	Brawata	Techniques for monitoring carnivore behavior using automatic thermal video	Dingo, Fox (Red)	Camera	FLIR	ThermaCAM S45	Ground-based static
2013	Yang	Study of bat flight behavior by combining thermal image analysis with a LiDAR forest reconstruction	Bats	Camera	FLIR	ThermoVision SC8000	Ground-based static
2014	Beaver	Aerial vertical-looking infrared imagery to evaluate bias of distance sampling techniques for white-tailed deer	Deer	Thermal monocular	Thermal-Eye/ Mitsubishi	250D ProTech, Berea/IR-M500	Aerial (plane) and vehicle-mounted
2014	Christiansen	Automated detection and recognition of wildlife using thermal cameras	Rabbit and chicken	Camera	FLIR	Not specified	Fitted on lift

Year	Lead author	Title	Target species group	Device type	Make	Model	Deployment type
2014	Cryan	Behavior of bats at wind turbines	Bats	Camera	Axis	Q1921-E	Placed looking up towards sky on ground at base of wind turbines
2014	Mulero-Pazmany	Remotely piloted aircraft systems as a rhinoceros anti-poaching tool in Africa	Rhinoceros	Camera	Thermoteknix	Micro CAM	Drone-mounted
2014	Robinson	Factors influencing the efficacy of forward-looking infrared in Polar Bear den detection	Polar Bear	Camera	FLIR	ThermaCAM P65HS	Ground-based – roaming
2014	Wu	A thermal infrared video benchmark for visual analysis	Bats	Camera	FLIR	SC8000	Tripod, docked in corner of corridors and rooms
2015	Berthinussen	WC1060 Development of a cost-effective method for monitoring the effectiveness of mitigation for bats crossing linear transport infrastructure	Bats	Camera	FLIR	H Series	Tripod
2015	Chretien	Wildlife multispecies remote sensing using visible and thermal infrared imagery acquired from an unmanned aerial vehicle (UAV)	Bison, Deer (Fallow), Wolf (Gray), Elk (Incidental: Ostrich, Coyotes, Black Bears)	Uncooled VOx microbolometer	FLIR	Tau640	Fixed to UAV
2015	Cullinan	Classification of birds and bats using flight tracks	Birds (Gulls, Swallows), Bats	Camera	Axsys	FieldPro 5X	Fixed on Tripod
2015	Ganow	Use of thermal imaging to estimate the population sizes of Brazilian free-tailed bat, *Tadarida brasiliensis*, maternity roosts in Oklahoma	Bats	Camera	FLIR	SR-19	Ground-based – static
2015	Hart	Can handheld thermal imaging technology improve detection of poachers in African bushveldt?	Poachers	Camera	FLIR	T620; i7	Handheld

Year	Lead author	Title	Target species group	Device type	Make	Model	Deployment type
2015	Horton	A comparison of traffic estimates of nocturnal flying animals using radar, thermal imaging, and acoustic recording	Nocturnal birds	Camera	FLIR	Guardsman HG-307 Pro	Tripod
2015	Lhoest	How Many Hippos (Homhip): Algorithm for Automatic Counts of Animals with Infra-Red Thermal Imagery from UAV	Hippos	Camera	Tamarisk	640	Drone-mounted
2015	Matzner	Two-dimensional thermal video analysis of offshore bird and bat flight	Birds, Bats	Not specified	Not specified	Not specified	Not specified
2016	Austin	If waterbirds are nocturnal are we conserving the right habitats?	Waterbirds	Camera	Thermopro	TP8	Tripod-mounted
2016	Chandra Kumar	Averting Wildlife–Train Interaction Using Thermal Image Processing and Acoustic Techniques	Various – elephant, rhino, bee	Not specified	Not specified	Not specified	Not specified
2016	Diehl	Evaluating the effectiveness of wildlife detection and observation technologies at a solar power tower facility	Insects, Bats, Birds	Camera and Surveillance camera	FLIR/AXIS	SC8343HD/ Q1921-E	Ground-based static
2016	Flodell	Wildlife surveillance using a uav and thermal imagery	Multiple	Not specified			
2016	Gonzalez	Unmanned aerial vehicles (UAVs) and artificial intelligence revolutionizing wildlife monitoring and conservation	Multiple (focuses on Koala)	Drone camera	FLIR	Tau 2 640	Drone-mounted
2016	Gooday	An assessment of thermal-image acquisition with an unmanned aerial vehicle (UAV) for direct counts of coastal marine mammals ashore	Seals	Camera (2)	Unspecified and Optris	T320 and PL450	Drone-mounted and ground-based
2016	Hampton	An assessment of animal welfare for the culling of peri-urban kangaroos	Kangaroo	Scope	Guide	518C	Not specified
2016	Kano	Nasal temperature drop in response to a playback of conspecific fights in chimpanzees: a thermos-imaging study	Chimpanzee	Camera	FLIR	T650sc	

Year	Lead author	Title	Target species group	Device type	Make	Model	Deployment type
2016	Mammeri	Animal–vehicle collision mitigation system for automated vehicles	Moose and Cervidae	N/A	N/A	N/A	
2016	Ward	Autonomous UAVs wildlife detection using thermal imaging, predictive navigation and computer vision	Various (tested on dog)	Sensor	FLIR	Lepton	Drone-mounted
2017	Bartonička	Deeply torpid bats can change position without elevation of body temperature	Bats (Greater Mouse-eared Bat)	Camera	Guide	M8	Not specified
2017	Dezecache	Skin temperature and reproductive condition in wild female chimpanzees	Chimpanzee	Not specified	Testo	881–2	Not specified
2017	Dolan	Techniques: Thermal birding	Birds	Camera	Pulsar	XD50S	Ground-based – roaming
2017	Hayman	Long-term video surveillance and automated analyses reveal arousal patterns in groups of hibernating bats	Bats (Little Brown Bat, Indiana Bat)	Surveillance cameras	Axis	Q-1921E	Ground-based – static
2017	Johnston	Comparing occupied and unoccupied aircraft surveys of wildlife populations: assessing the gray seal (*Halichoerus grypus*) breeding colony on Muskeget Island, USA	Grey Seals	Drone camera	senseFly	Thermomap 1.2MP	Drone-mounted
2017	Longmore	Adapting astronomical source detection software to help detect animals in thermal images obtained by unmanned aerial systems	Tested on cows	Drone camera	Not specified	Not specified	Drone-mounted
2017	Seymour	Automated detection and enumeration of marine wildlife using unmanned aircraft systems (UAS) and thermal imagery	Grey Seals	Drone camera	senseFly	Thermomapper	Drone-mounted
2018	Dahlen	Successful aerial survey using thermal camera to detect wild orangutans in a fragmented landscape	Orang-utan	Drone camera (2)	FLIR	FLIR™ 640 Pro and Zenmuse XT 640 × 512	Drone-mounted
2018	Faber Johannesen	The Efficiency of Thermal Imaging for Wildlife Monitoring	Various	Camera	FLIR	T1020	Tripod-mounted

Year	Lead author	Title	Target species group	Device type	Make	Model	Deployment type
2018	Goodenough	Identification of African Antelope Species: Using Thermographic Videos to Test the Efficacy of Real-Time Thermography	Antelope	Camera	FLIR	T620	Vehicle-mounted
2018	Kinzie	Ultrasonic Bat Deterrent Technology	Bats	Camera	AXIS	Q1922-E and Q1932_E	Ground-based – static
2018	Oishi	Animal Detection Using Thermal Images and Its Required Observation Conditions	Various (human, dog, Sika Deer)	Camera and sensor	NEC Avio/ Nakanihon Air Service Co	ThermoShot F30/ITRES TABI-1800	Ground-based – static (BRIDGE) and Aerial
2018	Witczuk	Exploring the feasibility of unmanned aerial vehicles and thermal imaging for ungulate surveys in forests-preliminary results	Ungulates (Red Deer, Roe Deer, Wild Boar)	Camera	Vigo	IRMOD v640	Drone-mounted
2019	Abdul-Mutalib	Feasibility of thermal imaging using unmanned aerial vehicles to detect Bornean orangutans	Orang-utan	Thermal camera attached to drone	FLIR	Vue 640	Drone-mounted
2019	Bowen	An evaluation of thermal infrared cameras for surveying hedgehogs in parkland habitats	Hedgehogs	Handheld imager	FLIR	E60 TIC	Ground-based – roaming
2019	Burke	Optimizing observing strategies for monitoring animals using drone-mounted thermal infrared cameras	Riverine Rabbit	Drone camera	FLIR	T640 TIR	Drone-mounted
2019	Burke	Successful observation of orangutans in the wild with thermal-equipped drones	Orang-utan	Drone-mounted	Thermal Capture/FLIR	Fusion Zoom/ Tau 2 640	Drone-mounted
2019	Bushaw	Applications of unmanned aerial vehicles to survey mesocarnivores	Mesocarnivores (Red Fox, Coyote, Striped Skunk)	Drone camera	DJI	Zenmuse XT2 R	Drone-mounted
2019	Chretien	Visible and thermal infrared remote sensing for the detection of white-tailed deer using an unmanned aerial system	Deer (White-tailed)	Drone camera	FLIR	Tau640	Drone-mounted
2019	Gilmour	Evaluating methods to deter bats	Bats	Thermal camera	Optris	PI640	Ground-based – static

Year	Lead author	Title	Target species group	Device type	Make	Model	Deployment type
2019	Hambrecht	Detecting 'poachers' with drones: Factors influencing the probability of detection with TIR and RGB imaging in miombo woodlands, Tanzania	Poachers	Thermal cameras attached to drone	TeAx	Thermal Capture v1.0 TIR (TAU 640 core)	Drone-mounted
2019	Heintz	Exploratory investigation of infrared thermography for measuring gorilla emotional responses to interactions with familiar humans	Gorilla	Thermal camera	FLIR	T650sc	Ground-based – static
2019	Kays	Hot monkey, cold reality: surveying rainforest canopy mammals using drone-mounted thermal infrared sensors	Various (monkeys, sloths, anteater)	Thermal camera attached to drone	DJI	Zenmuse XT® V2.0 FLIR uncooled thermal infrared radiometric sensor	Drone-mounted
2019	Lethbridge	Estimating kangaroo density by aerial survey: a comparison of thermal cameras with human observers	Kangaroo	Thermal camera attached to drone	FLIR	T1K	Aerial (Helicopter)
2019	Mitchell	Using infrared thermography to detect night-roosting birds	Night-roosting birds	Camera	InfReC	G120EX	Ground-based – roaming
2019	Narayan	Using Thermal Imaging to Monitor Body Temperature of Koalas (Phascolarctos cinereus) in A Zoo Setting	Koala	Thermal camera	FLIR	530	Ground-based – handheld
2019	Nottingham	Snacks in the city: the diet of hedgehogs in Auckland urban forest fragments	Hedgehogs	Thermal scope	Pulsar	Quantum XD19	Ground-based – roaming
2019	Otálora-Ardila	Thermally-assisted monitoring of bat abundance in an exceptional cave in Brazil's Caatinga drylands	Bats	Handheld imager	FLIR	E60	Ground-based – static
2019	Roberts	Testing the efficacy of a thermal camera as a search tool for locating wild bumble bee nests	Bumblebee nests	Handheld imager	FLIR	E60	Ground-based – roaming

Year	Lead author	Title	Target species group	Device type	Make	Model	Deployment type
2019	Scholten	Real-time thermal imagery from an unmanned aerial vehicle can locate ground nests of a grassland songbird at rates similar to traditional methods	Songbirds	Drone camera	FLIR	XT Zenmuse	Drone-mounted
2019	Semel	Eyes in the sky: Assessing the feasibility of low-cost, ready-to-use Unmanned Aerial Vehicles to monitor primate populations directly	Sifaka	Drone camera	FLIR	Not specified	Drone-mounted
2019	Spaan	Thermal infrared imaging from drones offers a major advance for spider monkey surveys	Spider Monkey	Drone camera	FLIR	Tau2 640	Drone-mounted
2019	Stephenson	Quantifying thermal-imager effectiveness for detecting bird nests on farms	Birds	Camera	FLIR	E8	Ground-based – roaming
2020	Anderson	Deviation Factors in the Mississippi Flyway: Geographic Barriers and Ecological Quality	Passerines	Security camera	FLIR	SR-19	Ground-based – static
2020	Atuchin	Drone-Assisted Aerial Surveys of Large Animals in Siberian Winter Forests	European Elk	Drone camera	ATOM	M500	Drone-mounted
2020	Beaver	Evaluating the Use of Drones Equipped with Thermal Sensors as an Effective Method for Estimating Wildlife	White-tailed Deer	Drone camera	FLIR	Vue Pro 640	Drone-mounted
2020	Burger	Nightly selection of resting sites and group behavior reveal antipredator strategies in giraffe	Giraffe	Thermal camera	SEEK	CompactPro	Vehicle-mounted
2020	Chapman	Anti-poaching strategies employed by private rhino owners in South Africa	Rhinoceros	N/A	N/A	N/A	N/A
2020	Corcoran	New technologies in the mix: Assessing N-mixture models for abundance estimation using automated detection data from drone surveys	Koala	Drone camera	FLIR	Tau 2 640	Drone-mounted
2020	Croft	Too many wild boar? Modelling fertility control and culling to reduce wild boar numbers in isolated populations	Wild Boar	Not specified	Not specified	Not specified	Not specified

Year	Lead author	Title	Target species group	Device type	Make	Model	Deployment type
2020	da Costa	Diagnostic Applications of Infrared Thermography in Captive Brazilian Canids and Felids	Canids and Felids	Thermal camera	FLIR	T640	Ground-based – static
2020	de Quieros	How does Spix's yellow-toothed cavy (*Galea spixii* Wagler, 1831) face the thermal challenges of the Brazilian tropical dry forest?	Yellow-toothed Cavy	Handheld imager	FLIR	b60	Handheld
2020	Fletcher	Observations on breeding and dispersal by Capercaillie in Strathspey	Capercaillie	Scope	Pulsar	Quantum XP50	Ground-based – roaming
2020	Graveley	Using a thermal camera to measure heat loss through bird feather coats	Birds	Thermal camera	FLIR	SC655	Ground-based – static
2020	Hamilton	Automated Detection of Koalas on Kangaroo Island, South Australia	Koalas	Drone camera	Not specified	Not specified	Drone-mounted
2020	Helvey	Duck Nest Detection Through Remote Sensing	Ducks	Drone camera	Tamarisk	640 Longwave Infrared	Drone-mounted
2020	Hyun	Remotely piloted aircraft system (Rpas)-based wildlife detection: a review and case studies in maritime Antarctica	Southern Elephant Seal	Drone camera	FLIR	Vue Pro R	Drone-mounted
2020	Karp	Detecting small and cryptic animals by combining thermography and a wildlife detection dog	Brown Hare	Scope	FLIR	Scout TS-32r Pro	Vehicle-mounted
2020	Kilgo	Use of roadside deer removal to reduce deer–vehicle collisions	Deer (White-tailed)	Scope	Not specified	Not specified	Not specified
2020	Jumail	A comparative evaluation of thermal camera and visual counting methods for primate census in a riparian forest at the Lower Kinabatangan Wildlife Sanctuary (LKWS), Malaysian Borneo	Primate census	Drone camera	FLIR	Tau 2 640	Boat
2020	Landry	Interactions between livestock guarding dogs and wolves in the southern French Alps	Wolves	Binoculars	MATIS	Not specified	Ground-based – static

Year	Lead author	Title	Target species group	Device type	Make	Model	Deployment type
2020	Lethbridge	Report of state-wide census of wild fallow deer in Tasmania project: Part A: Baseline aerial survey of fallow deer population, central and north-eastern Tasmania	Fallow Deer	Drone camera	FLIR	T1K	Aerial (Helicopter)
2020	Martin	Thermal biology and growth of bison (*Bison bison*) along the Great Plains: examining four theories of endotherm body size	Bison	Thermal camera	FLIR	T1030sc	Not specified
2020	Prosekov	Not available	European Moose, Roe Deer, Wolf, Brown Bear	Not available	Not available	Not available	Not available
2020	Rahman	An experimental approach to exploring the feasibility of unmanned aerial vehicle and thermal imaging in terrestrial and arboreal mammals research	Javan Deer, Long-tailed Macaque	Drone camera	FLIR	Not specified	Drone-mounted
2020	Santangeli	Integrating drone-borne thermal imaging with artificial intelligence to locate bird nests on agricultural land	Birds	Drone camera	FLIR/DJI	Tau 2	Drone-mounted
2020	Schedl	Airborne optical sectioning for nesting observation	Heron	Drone camera	FLIR	Vue Pro	Drone-mounted
2020	Schirmacher	Evaluating the effectiveness of an ultrasonic acoustic deterrent in reducing bat fatalities at wind energy facilities	Bats	Thermal camera	AXIS	Q1932-e	Ground-based – static
2020	Shewring	Using UAV-mounted thermal cameras to detect the presence of nesting nightjar in upland clear-fell: A case study in South Wales, UK	Nightjar	Thermal camera attached to drone	Falcon/FLIR/DJI/ FLIR	8/Tau 2/T600 Inspire 1/ Zenmuse XT V2.0	Drone-mounted
2020	Shirai	The Effectiveness of Visual Scaring Techniques Against Grey Herons, *Ardea cinerea*	Grey Heron	Thermal camera	FLIR	Scout TK	Not specified
2020	Simmons	Big brown bats are challenged by acoustically-guided flights through a circular tunnel of hoops	Big Brown Bat	Thermal camera	Photon	320	Not specified
2020	Smallwood	Effects of Wind Turbine Curtailment on Bird and Bat Fatalities	Bats and birds	Thermal camera	FLIR	T620	Not specified

Year	Lead author	Title	Target species group	Device type	Make	Model	Deployment type
2020	Smallwood	Relating bat passage rates to wind turbine fatalities	Bats	Thermal camera	FLIR	T620	Not specified
2020	Smallwood	Dogs detect larger wind energy effects on bats and birds	Bats and birds	Not specified	Not specified	Not specified	Not specified
2020	Smith	Efficacy of aerial forward-looking infrared surveys for detecting Polar Bear maternal dens	Polar Bear	Camera	FLIR	Star Safire (models II, III, and HD 380)	Aerial (Plane)
2020	South	Diagnosis of hypothermia in the European hedgehog, *Erinaceus europaeus*, using infrared thermography	Hedgehog	Handheld imager	FLIR	E60bx	Not specified
2020	Vinson	Thermal cameras as a survey method for Australian arboreal mammals: a focus on the greater glider	Arboreal mammals (Greater Glider)	Handheld imager and thermal camera	FLIR	ThermaCAM S65/T1050sc	Ground-based – roaming
2020	Voigt	Survival rates on pre-weaning European hares (*Lepus europaeus*) in an intensively used agricultural area	Hare (European)	Camera	Raytheon	Palm IR 250-D	Ground-based – roaming
2020	West Midlands Ringing Group	Getting The Most From Your Thermal Camera A Guide To Using A Thermal Image Camera For Ringing And Surveying	Skylark, Redwing, Fieldfare, Jack Snipe, Yellowhammer	Thermal scope	Pulsar Helion XQ	28 mm; 38 mm; 50 mm	Handheld
2020	Witt	Real-time drone derived thermal imagery outperforms traditional survey methods for an arboreal forest mammal	Koala	Drone camera	DJI	Matrice 200 v2	Drone-mounted
2020	Zainalabidin	Distribution of arboreal nocturnal mammals in northern Borneo	Nocturnal arboreal mammals	Scope	FLIR	Scout III 640	Ground-based – roaming
2021	Ampeng	First Bornean orangutan sighting in Usun Apau National Park, Sarawak	Orang-utan	Drone Camera	FLIR	Not specified	Drone-mounted

Year	Lead author	Title	Target species group	Device type	Make	Model	Deployment type
2021	Anderson	Flight directions of songbirds are unaffected by the topography of Lake Erie's southern coastline during fall migration	Songbirds	Camera (Security)	FLIR	SR-19	Ground-based – roaming
2021	Baker	Infrared antenna-like structures in mammalian fur	Mouse, Antechinus, Shrew, Rabbit	Camera	Leonardo	Merlin	Not specified
2021	Bedson	Estimating density of mountain hares using distance sampling: a comparison of daylight visual surveys, night-time thermal imaging and camera traps	Hare (Mountain)	Binocular	Armasight	Command 336 HD	Ground-based – roaming
2021	Boczkowski	Analysis of the Possibility of Using Unmanned Aerial Vehicles and Thermovision for the Stocktaking of Big Game	Moose, Deer (Fallow, Roe), Boar, Mouflon, Fox, Racoon Dog, Badger, Marten, Hare, Pheasant	Scopes and Smartphone attachment and Drone-mounted Camera (and Binocular??)	Pulsar Pulsar HKVision Seek	Thermion XP50 Accolade 2 XP50 LRF Pro Lynx LH15 Thermal Compact Pro	All handheld
2021	Bushaw	Application of unmanned aerial vehicles and thermal imaging cameras to conduct duck brood surveys	Duck	Drone Camera	FLIR	Zenmuse XT	Drone-mounted
2021	Corcoran, A. J.	ThruTracker: open-source software for 2-D and 3-D animal video tracking	Bats, Birds	Drone Camera	FLIR	Zenmuse XT and A65	Not specified
2021	Corcoran, E.	Evaluating new technology for biodiversity monitoring: Are drone surveys biased?	Koala	Drone Camera	FLIR	Tau 2 640	Drone-mounted
2021	Cox	Hot stuff in the bushes: Thermal imagers and the detection of burrows in vegetated sites	Rabbit	Drone Cameras (3)	FLIR/Jenoptik/ Sierra-Olympic	Zenmuse XT 640/VarioCAM HD/Vayu HD	Drone-mounted
2021	Filipovs	UAV areal imagery-based wild animal detection for sustainable wildlife management	Elk, Deer (Red, Roe), Boar	Drone Camera	Not specified	Not specified	Drone-mounted
2021	Florko	Narwhal (*Monodon monoceros*) detection by infrared flukeprints from aerial survey imagery	Narwhal	Camera	FLIR	T1020sc	Aerial (Plane)

Year	Lead author	Title	Target species group	Device type	Make	Model	Deployment type
2021	Gurnell	Surveys of Hedgehogs in The Regent's Park, London	Hedgehog	Camera	FLIR	E60	Ground-based – roaming
2021	Hohmann	The possibilities and limitations of thermal imaging to detect wild boar (*Sus scrofa*) carcasses as a strategy for managing African Swine Fever (ASF) outbreaks	Boar	Multiple:	FLIR	A65 and Star SAFIRE 380HDC	Ground-based – static and Aerial (Helicopter)
2021	Kim	A Manual for Monitoring Wild Boars (*Sus scrofa*) Using Thermal Infrared Cameras Mounted on an Unmanned Aerial Vehicle (UAV)	Boar	Drone Camera	FLIR	Zenmuse XT2	Drone-mounted
2021	Lee	Feasibility Analyses of Real-Time Detection of Wildlife Using UAV-Derived Thermal and RGB Images	Modelled on alpacas	Drone Camera	FLIR	Zenmuse XT2	Drone-mounted
2021	McCarthy	Drone-based thermal remote sensing provides an effective new tool for monitoring the abundance of roosting fruit bats	Fruit bats	Drone Camera	FLIR	Zenmuse XT	Drone-mounted
2021	McGregor	Effectiveness of thermal cameras compared to spotlights for counts of arid zone mammals across a range of ambient temperatures	Rodents, Bettongs, Bilbies, Rabbits	Scope	FLIR	Scout III 640	Vehicle-based
2021	McMahon	Evaluating Unmanned Aerial Systems for the Detection and Monitoring of Moose in Northeastern Minnesota	Moose	Drone Camera	FLIR	Vue Pro and Duo Pro	Drone-mounted
2021	Medolago	Use of a portable thermograph as a potential tool to identify nocturnal airport bird risks	Birds	Camera	FLIR	T650sc	Ground-based – roaming
2021	Playà-Montmany	Spot size, distance and emissivity errors in field applications of infrared thermography	Trumpeter Swan	Cameras	FLIR	SC660; T1030sc; 1030sc	Not specified
2021	Preston	Enumerating White-Tailed Deer Using Unmanned Aerial Vehicles	White-tailed Deer	Drone Camera	FLIR	Vue Pro R	Drone-mounted
2021	Psiroukis	Monitoring of free-range rabbits using aerial thermal imaging	Rabbits	Drone Camera	FLIR	Vue Pro	Drone-mounted

Year	Lead author	Title	Target species group	Device type	Make	Model	Deployment type
2021	Rahman	Performance of unmanned aerial vehicle with thermal imaging, camera trap, and transect survey for monitoring of wildlife	Deer, Shrew, Eagle	Drone Camera	FLIR	Vue Pro	Drone-mounted
2021	Sliwinski	Comparison of spotlighting monitoring data of European brown hare (*Lepus europaeus*) relative population densities with infrared thermography in agricultural landscapes in Northern Germany	Hare (European Brown)	Camera?	Nyxus	Bird	Ground-based – roaming
2021	Syposz	Avoidance of different durations, colours and intensities of artificial light by adult seabirds	Manx Shearwaters	Camera	FLIR	T640	Ground-based – static and Aerial (Helicopter)
2021	Zini	Human and environmental associates of local species-specific abundance in a multi-species deer assemblage	Deer (Roe, Muntjac)	Scope	Pulsar	Helion XP50	Vehicle-based

Table 2 Key information from literature including the use of thermal imaging for terrestrial mammals between 1968 and 2021 (inclusive).

Year	Lead author	Title	Device type	Make	Model	Resolution	Deployment type
1968	Croon	Infrared Scanning Techniques For Big Game	Line Scanner	Stoll-Hardy	Radiometer	N/A	Aerial (Plane)
1972	Graves	Censusing white-tailed deer by airborne thermal infrared imagery	Line Scanner	Not specified	Detector (In:Sb)/ MMercury Cadmium Telleride (MCT)	N/A	Aerial (Plane)
1972	Parker	An Experiment in Deer Detection by Thermal Scanning	Scanner	Bendix	LN-2-LW	N/A	Aerial (Plane)
1973	Cena	Thermographic measurements of the surface temperatures of animals	Camera	Agavision System	680	N/A	Vehicle-mounted
1977	Wride	Thermal Imagery for Census of Ungulates	Line Scanner	Daedalus	Not specified	N/A	Aerial (Plane)

Year	Lead author	Title	Device type	Make	Model	Resolution	Deployment type
1992	Klir	An infrared thermographic study of surface temperature in relation to external thermal stress in three species of foxes: The red fox (*Vulpes vulpes*), arctic fox (*Alopex lagopus*), and kit fox (*Vulpes macrotis*)	Camera/Scanner	Inframetrics	525	Not specified	Ground-based – static
1993	Wiggers	Use of thermal infrared sensing to survey white-tailed deer populations	Scanner	FLIR	2000G	Not specified	Aerial (Plane)
1994	Boonstra	Finding mammals using far-infrared thermal imaging	Not specified	Agema	Thermovision 210	Not specified	Ground-based – not specified
1995	Garner	Use of modern infrared thermography for wildlife population surveys	Scanner	FLIR	2000 A/B	Not specified	Aerial (Plane)
1995	Havens	The use of thermal imagery in the aerial survey of panthers (and other animals) in the Florida Panther National Wildlife Refuge and the Big Cypress National Preserve	Camera	InfraCam	Not specified	Not specified	Aerial (Plane)
1996	Naugle	Use of thermal infrared sensing to estimate density of white-tailed deer	Not specified	FLIR	2000AB	Not specified	Aerial (Plane)
1997	Gill	The use of portable thermal imaging for estimating deer population density in forest habitats	Weapons targeting device	Pilkington Thorn Optronics	LITE direct view thermal imager	Not specified	Vehicle-mounted
1998	Havens	Using thermal imagery in the aerial survey of animals	Not available	Not available	Not available	Not available	Not available
1999	Haroldson	Evaluation of thermal infrared imaging for detection of white-tailed deer	Not available	Not available	Not available	Not available	Not available
2000	Belant	Comparison of 3 devices to observe white-tailed deer at night	Not available	Not available	Not available	Not available	Not available
2002	Arenas	An evaluation of the application of infrared thermal imaging to the tele-diagnosis of sarcoptic mange in the Spanish ibex (*Capra pyrenaica*)	Camera	Elbit	Milcam	Not specified	Ground-based

Year	Lead author	Title	Device type	Make	Model	Resolution	Deployment type
2002	Dunn	Using thermal infrared sensing to count elk in the south-western United States	Not specified	Inframetrics	IRTV-445G MK II FLIR	Not specified	Aerial (Aeroplane)
2003	Haroldson	Evaluation of aerial thermal imaging for detecting white-tailed deer in a deciduous forest environment	Not specified	Westinghouse Electronic Systems	WF-160DS	Not specified	Aerial (Aeroplane)
2004	Bernatas	Sightability model for California bighorn sheep in canyonlands using forward-looking infrared (FLIR)	Camera	FLIR	Westinghouse WesCam DS16 FLIR	Not specified	Aerial (Aeroplane)
2004	Kissell	An assessment of thermal infrared detection rates using white-tailed deer surrogates	Not specified	Mitsubishi	IR-M700	Not specified	Aerial (Aeroplane)
2004	Smart	Monitoring woodland deer populations in the UK: an imprecise science	Weapons targeting device	Pilkington Thorn Optronics	LITE direct view thermal imager	Not specified	Vehicle-mounted
2005	Ditchkoff	From the Field: capture of white-tailed deer fawns using thermal imaging technology	Camera	Raytheon	Palm IR 250 Digital	Not specified	Vehicle-mounted
2005	Drake	Counting a suburban deer population using forward looking infrared radar and road counts	Camera	FLIR	Series 2000F	Not specified	Vehicle-mounted and Aerial (Helicopter)
2005	Hodnett	Thermal imaging applications in urban deer control	Camera	Wescam	DS200	Not specified	Aerial (Helicopter mounted)
2005	Lavers	Application of remote thermal imaging and night vision technology to improve endangered wildlife resource management with minimal animal distress and hazard to humans	Camera	Raytheon	PalmIR Pro	320 × 256	Tripod scaled with normal cameras on same target
2005	Potvin	Testing 2 aerial survey techniques on deer in fenced enclosures: visual double-counts and thermal infrared sensing	Camera	FLIR	2000 A/B sensor	Not specified	Aerial (Fixed to helicopter)
2006	Butler	Limitations of thermal infrared imaging for locating neonatal deer in semiarid shrub communities	Camera	Raytheon	PalmIR 250 Digital	Not specified	Tripod fixed on pickup cab

Year	Lead author	Title	Device type	Make	Model	Resolution	Deployment type
2006	Dunbar	Use of infrared thermography to detect signs of rabies infection in raccoons (*Procyon lotor*)	Camera	FLIR	ThermaCAM E-65	Not specified	Not specified
2007	Hemami	Estimating abundance of introduced Chinese muntjac *Muntiacus reevesi* and native roe deer *Capreolus capreolus* using portable thermal imaging equipment	Weapons targeting device	Pilkington Thorn Optronics	LITE direct view thermal imager	Not specified	Vehicle-mounted
2007	Sumbera	Patterns of surface temperatures in two male-rats (Bathyergidae) with different social system as revealed by IR-thermography	Camera	Not specified	AGA 570	Not specified	Not specified
2009	Dunbar	Use of infrared thermography to detect thermographic change in mule deer (*Odocoileus hemionus*) experimentally infected with foot-and-mouth disease	Camera	FLIR	ThermaCAM EX320	Not specified	Not specified
2009	Kuhn	Infrared thermography of the body surface in the Eurasian otter *Lutra lutra* and the giant otter *Pteronura brasiliensis*	Camera	FLIR	ThermaCAM B20	Not specified	Not specified
2009	Lavers	Non-destructive high-resolution thermal imaging techniques to evaluate wildlife and delicate biological samples	Camera	FLIR; Echotherm	ThermaCAM E320; Thermal Wave Imaging	320 × 240; 640 × 512	Not specified
2009	Marini	Response to human presence during nocturnal line transect surveys in fallow deer (*Dama dama*) and wild boar (*Sus scrofa*)	Camera	FLIR	ThermaCAM™ PM545	320 × 240	Not specified
2010	Carr	Comparative woodland caribou population surveys in Slate Islands Provincial Park, Ontario	Camera	Agema	Thermovision 1000	800 × 400	Aerial (Plane)
2010	Weissenböck	Thermal windows on the body surface of African elephants (*Loxodonta africana*) studied by infrared thermography	Camera	FLIR	ThermaCAM P60	Not specified	Not specified
2011	Israel	A UAV-based roe deer fawn detection system	Camera	FLIR	Tau 640	640 × 512	Aerial (Drone)
2011	Kissell	A technique to estimate white-tailed deer density using vertical-looking infrared imagery	Camera	Mitsubishi	IR-M700	Not specified	Aerial (Plane)

Year	Lead author	Title	Device type	Make	Model	Resolution	Deployment type
2011	Millette	AIMS-Thermal – a Thermal and High Resolution Color Camera System Integrated with GIS for Aerial Moose and Deer Census in Northeastern Vermont	Camera system	AIMS	Not specified	640 × 480	Aerial (Plane)
2011	Storm	Comparison of visual-based helicopter and fixed-wing forward looking infrared surveys for counting white-tailed deer *Odocoileus*	Camera	PolyTech	Kelvin 350 II infrared sensor	Not specified	Aerial (Plane)
2012	Franke	Aerial ungulate surveys with a combination of infrared and high-resolution natural colour images	Camera	JENOPTIK	Not specified	640 × 480	Aircraft-mounted (Airplane)
2012	Zhou	Contour-based HOG deer detection in thermal images for traffic safety	Camera	Thermal-Eye	TSCss-FF	Not specified	Fitted on rotating motor. 180 range
2012	Morelle	Game species monitoring using road-based distance sampling in association with thermal imagers: a covariate analysis	Scope/Camera	FLIR/JENOPTIK	ThermaCAM™ HS–324/ VarioCAM™	320 × 24/640 × 480	Handheld in car
2012	Steen	Automatic Detection of Animals in Mowing Operations Using Thermal Cameras	Camera	FLIR	Not specified	Not specified	Fitted on mowing tractor
2012	Franzetti	Nocturnal line transect sampling of wild boar (*Sus scrofa*) in a Mediterranean forest: long-term comparison with capture-mark-resight population estimates	Camera	FLIR	ThermaCAM PM545® and ThermaCAM™ B640	Not specified	Ground-based – roaming
2013	Department for Environment Food and Rural Affairs	Monitoring the humaneness of badger population reduction by controlled shooting	Not specified	Not specified	Not specified	Not specified	Ground-based – roaming
2013	Brawata	Techniques for monitoring carnivore behavior using automatic thermal video	Camera	FLIR	ThermaCAM S45	Not specified	Ground-based static

Year	Lead author	Title	Device type	Make	Model	Resolution	Deployment type
2014	Beaver	Aerial vertical-looking infrared imagery to evaluate bias of distance sampling techniques for white-tailed deer	Thermal scope	Thermal-Eye/ Mitsubishi	250D ProTech, Berea/IR-M500	Not specified	Aerial (plane) and vehicle-mounted
2014	Christiansen	Automated detection and recognition of wildlife using thermal cameras	Camera	FLIR	Not specified	320 × 240	Fitted on lift
2014	Mulero-Pazmany	Remotely piloted aircraft systems as a rhinoceros anti-poaching tool in Africa	Camera	Thermoteknix	Micro CAM	640 × 480	Drone-mounted
2015	Chretien	Wildlife multispecies remote sensing using visible and thermal infrared imagery acquired from an unmanned aerial vehicle (UAV)	Uncooled VOx microbolometer	FLIR	Tau640	640 × 480	Fixed to UAV
2015	Lhoest	How Many Hippos (Homhip): Algorithm for Automatic Counts of Animals with Infra-Red Thermal Imagery from UAV	Camera	Tamarisk	640	640 × 480	Drone-mounted
2016	Chandra Kumar	Averting Wildlife–Train Interaction Using Thermal Image Processing and Acoustic Techniques	Not specified	Not specified	Not specified	Not specified	Not specified
2016	Mammeri	Animal–vehicle collision mitigation system for automated vehicles	N/A	N/A	N/A	N/A	N/A
2016	Ward	Autonomous UAVs wildlife detection using thermal imaging, predictive navigation and computer vision	Sensor	FLIR	Lepton	60 × 80	Drone-mounted
2017	Longmore	Adapting astronomical source detection software to help detect animals in thermal images obtained by unmanned aerial systems	Drone camera	Not specified	Not specified	Not specified	Drone-mounted
2018	Faber Johannesen	The Efficiency of Thermal Imaging for Wildlife Monitoring	Camera	FLIR	T1020	1024 × 768	Tripod-mounted
2018	Goodenough	Identification of African Antelope Species: Using Thermographic Videos to Test the Efficacy of Real-Time Thermography	Camera	FLIR	T620	640 × 480	Vehicle-mounted

Year	Lead author	Title	Device type	Make	Model	Resolution	Deployment type
2018	Oishi	Animal Detection Using Thermal Images and Its Required Observation Conditions	Camera and sensor	NEC Avio/ Nakanihon Air Service Co	ThermoShot F30/ITRES TABI-1800	Not specified	Ground-based – static (BRIDGE) and Aerial
2018	Witczuk	Exploring the feasibility of unmanned aerial vehicles and thermal imaging for ungulate surveys in forests-preliminary results	Camera	Vigo	IRMOD v640	640 × 480	Drone-mounted
2019	Bowen	An evaluation of thermal infrared cameras for surveying hedgehogs in parkland habitats	Handheld imager	FLIR	E60 TIC	Not specified	Ground-based – roaming
2019	Burke	Optimizing observing strategies for monitoring animals using drone-mounted thermal infrared cameras	Drone camera	FLIR	T640 TIR	640 × 512	Drone-mounted
2019	Burke	Successful observation of orangutans in the wild with thermal-equipped drones	Drone camera	Thermal Capture/FLIR	Fusion Zoom/ Tau 2 640	640 × 512/1920 × 1080	Drone-mounted
2019	Bushaw	Applications of unmanned aerial vehicles to survey mesocarnivores	Drone camera	DJI	Zenmuse XT2 R	640 × 512	Drone-mounted
2019	Chretien	Visible and thermal infrared remote sensing for the detection of white-tailed deer using an unmanned aerial system	Drone camera	FLIR	Tau640	Not specified	Drone-mounted
2019	Kays	Hot monkey, cold reality: surveying rainforest canopy mammals using drone-mounted thermal infrared sensors	Thermal camera attached to drone	DJI	Zenmuse XT® V2.0 FLIR uncooled thermal infrared radiometric sensor	640 × 512	Drone-mounted
2019	Nottingham	Snacks in the city: the diet of hedgehogs in Auckland urban forest fragments	Thermal scope	Pulsar	Quantum XD19	Not specified	Ground-based – roaming
2019	Spaan	Thermal infrared imaging from drones offers a major advance for spider monkey surveys	Drone camera	FLIR	Tau2 640	640 × 512	Drone-mounted

Year	Lead author	Title	Device type	Make	Model	Resolution	Deployment type
2020	Atuchin	Drone-Assisted Aerial Surveys of Large Animals in Siberian Winter Forests	Drone camera	ATOM	M500	35.9 × 24.0 mm/ 35 mm full frame	Drone-mounted
2020	Beaver	Evaluating the Use of Drones Equipped with Thermal Sensors as an Effective Method for Estimating Wildlife	Drone camera	FLIR	Vue Pro 640	640 × 480	Drone-mounted
2020	Burger	Nightly selection of resting sites and group behavior reveal antipredator strategies in giraffe	Thermal camera	SEEK	CompactPro	320 × 240	Vehicle-mounted
2020	Chapman	Anti-poaching strategies employed by private rhino owners in South Africa	N/A	N/A	N/A	N/A	N/A
2020	Croft	Too many wild boar? Modelling fertility control and culling to reduce wild boar numbers in isolated populations	Not specified	Not specified	Not specified	Not specified	Not specified
2020	da Costa	Diagnostic Applications Of Infrared Thermography In Captive Brazilian Canids And Felids	Thermal camera	FLIR	T640	640 × 480	Ground-based – static
2020	de Quieros	How does Spix's yellow-toothed cavy (*Galea spixii Wagler, 1831*) face the thermal challenges of the Brazilian tropical dry forest?	Handheld imager	FLIR	b60	Not specified	Handheld
2020	Karp	Detecting small and cryptic animals by combining thermography and a wildlife detection dog	Scope	FLIR	Scout TS-32r Pro	320 × 240	Vehicle-mounted
2020	Kilgo	Use of roadside deer removal to reduce deer–vehicle collisions	Scope	Not specified	Not specified	Not specified	Not specified
2020	Lethbridge	Report of state-wide census of wild fallow deer in Tasmania project: Part A: Baseline aerial survey of fallow deer population, central and north-eastern Tasmania	Drone camera	FLIR	T1K	Not specified	Aerial (Helicopter)
2020	Martin	Thermal biology and growth of bison (*Bison bison*) along the Great Plains: examining four theories of endotherm body size	Thermal camera	FLIR	T1030sc	1024 × 768	Not specified
2020	Prosekov	Not available	Not available	Not available	Not available	Not available	Not available

Year	Lead author	Title	Device type	Make	Model	Resolution	Deployment type
2020	Rahman	An experimental approach to exploring the feasibility of unmanned aerial vehicle and thermal imaging in terrestrial and arboreal mammals research	Drone camera	FLIR	Not specified	Not specified	Drone-mounted
2020	South	Diagnosis of hypothermia in the European hedgehog, *Erinaceus europaeus*, using infrared thermography	Handheld imager	FLIR	E60bx	Not specified	Not specified
2020	Voigt	Survival rates on pre-weaning European hares (*Lepus europaeus*) in an intensively used agricultural area	Camera	Raytheon	Palm IR 250-D	320 × 240	Ground-based – roaming
2020	Zainalabidin (inc Miard)	Distribution of arboreal nocturnal mammals in northern Borneo	Scope	FLIR	Scout III 640	Not specified	Ground-based – roaming
2021	Baker	Infrared antenna-like structures in mammalian fur	Camera	Leonardo	Merlin	Not specified	Not specified
2021	Bedson	Estimating density of mountain hares using distance sampling: a comparison of daylight visual surveys, night-time thermal imaging and camera traps	Binocular	Armasight	Command 336 HD	336 x 256	Ground-based – roaming
2021	Boczkowski	Analysis of the Possibility of Using Unmanned Aerial Vehicles and Thermovision for the Stocktaking of Big Game	Scopes and smartphone attachment and drone-mounted camera	Pulsar Pulsar HKVision Seek	Thermion XP50 Accolade 2 XP50 LRF Pro Lynx LH15 Thermal Compact Pro	640 × 480 640 × 480 384 × 284 384 × 284	All handheld
2021	Cox	Hot stuff in the bushes: Thermal imagers and the detection of burrows in vegetated sites	Drone cameras (3)	FLIR/Jenoptik/ Sierra-Olympic	Zenmuse XT 640/VarioCAM HD/Vayu HD	640 × 512/1024 × 800/1920 × 1200	Drone-mounted
2021	Filipovs	UAV areal imagery-based wild animal detection for sustainable wildlife management	Drone camera	Not specified	Not specified	Not specified	Drone-mounted
2021	Gurnell	Surveys of Hedgehogs in The Regent's Park, London	Camera	FLIR	E60	Not specified	Ground-based – roaming

Year	Lead author	Title	Device type	Make	Model	Resolution	Deployment type
2021	Hohmann	The possibilities and limitations of thermal imaging to detect wild boar (*Sus scrofa*) carcasses as a strategy for managing African Swine Fever (ASF) outbreaks	Multiple:	FLIR	A65 and Star SAFIRE 380HDC	Not specified	Ground-based – static and Aerial (Helicopter)
2021	Kim	A Manual for Monitoring Wild Boars (*Sus scrofa*) Using Thermal Infrared Cameras Mounted on an Unmanned Aerial Vehicle (UAV)	Drone camera	FLIR	Zenmuse XT2	640 × 12	Drone-mounted
2021	Lee	Feasibility Analyses of Real-Time Detection of Wildlife Using UAV-Derived Thermal and RGB Images	Drone camera	FLIR	Zenmuse XT2	640 × 512	Drone-mounted
2021	McGregor	Effectiveness of thermal cameras compared to spotlights for counts of arid zone mammals across a range of ambient temperatures	Scope	FLIR	Scout III 640	640 × 512	Vehicle-based
2021	McMahon	Evaluating Unmanned Aerial Systems for the Detection and Monitoring of Moose in Northeastern Minnesota	Drone camera	FLIR	Vue Pro and Duo Pro	640 × 512	Drone-mounted
2021	Preston	Enumerating White-Tailed Deer Using Unmanned Aerial Vehicles	Drone camera	FLIR	Vue Pro R	640 × 512	Drone-mounted
2021	Psiroukis	Monitoring of free-range rabbits using aerial thermal imaging	Drone camera	FLIR	Vue Pro	336 × 256	Drone-mounted
2021	Sliwinski	Comparison of spotlighting monitoring data of European brown hare (*Lepus europaeus*) relative population densities with infrared thermography in agricultural landscapes in Northern Germany	Monocular	Nyxus	Bird	Not specified	Ground-based – roaming
2021	Zini	Human and environmental associates of local species-specific abundance in a multi-species deer assemblage	Scope	Pulsar	Helion XP50	Not specified	Vehicle-based

Table 3 Literature including the use of thermal imaging for marine mammals between 1968 and 2021 (inclusive).

Year	Lead author	Title	Device type	Make	Model	Resolution	Deployment type
1972	Brooks	Infra-Red Scanning for Polar Bear	Scanner	Daedalus	Detector (In:Sb)/ Trimetal detector (Hg:Sb:Te)	N/A	Aerial (Plane)
1989	Barber	Thermal remote sensing for walrus population assessment in the Canadian arctic	Not specified	FLIR	Not specified	Not specified	Aerial (Plane)
1990	Kingsley	Infrared sensing of the under-snow lairs of the ringed seal	Infrared Sensor	FLIR	100A	Not specified	Aerial (Helicopter)
1991	Barber	Calibration of aerial thermal infrared imagery for walrus population assessment	Not specified	FLIR	Not specified	Not specified	Aerial (Plane)
1992	Cuyler	Thermal infrared radiation from free living whales	Not specified	Agema	Thermovision 880	Not specified	Boat
1999	Perryman	Diel variation in migration rates of eastern pacific gray whales measured with thermal imaging sensors	Not specified	Not specified	AN/KAS-1A	Not specified	Ground-based – static
2004	Armstrup	Detecting denning Polar Bears with forward looking infra-red (FLIR) imagery	Not available	Not available	Not available	Not available	Not available
2005	Willis	Spatial variation of heat flux in Steller sea lions: evidence for consistent avenues of heat exchange along the body trunk	Camera	FLIR	ThermaCAM PM695	Not specified	Ground-based – handheld
2006	Burn	Application of Airborne Thermal Imagery to Surveys of Pacific Walrus	Scanner	Daedalus	Airborne Multispectral Scanner	1489	Aircraft (Helicopter)
2008	Udevitz	Estimation of walrus populations on sea ice with infrared imagery and aerial photography	Scanner	Daedalus	Airborne Multispectral Scanner	1440	Aircraft (Helicopter)
2009	Burn	An improved procedure for detection and enumeration of walrus signatures in airborne thermal imagery	Scanner	Argon	Thermal infrared (8.5–12.5 mm) scanner	Not specified	Aircraft (Turbine aeroplane)

Year	Lead author	Title	Device type	Make	Model	Resolution	Deployment type
2010	Barbieri	Using infrared thermography to assess seasonal trends in dorsal fin surface temperatures of free-swimming bottlenose dolphins (*Tursiops truncatus*) in Sarasota Bay	Camera	FLIR	Agema 570 IR	Not specified	Monopod with regular camera on ship
2014	Robinson	Factors influencing the efficacy of forward-looking infrared in Polar Bear den detection	Camera	FLIR	ThermaCAM P65HS	Not specified	Ground-based – roaming
2016	Gooday	An assessment of thermal-image acquisition with an unmanned aerial vehicle (UAV) for direct counts of coastal marine mammals ashore	Camera (2)	Unspecified and Optris	T320 and PL450	Not specified	Drone-mounted and Ground-based
2017	Johnston	Comparing occupied and unoccupied aircraft surveys of wildlife populations: assessing the gray seal (*Halichoerus grypus*) breeding colony on Muskeget Island, USA	Drone camera	senseFly	Thermomap 1.2MP	Not specified	Drone-mounted
2017	Seymour	Automated detection and enumeration of marine wildlife using unmanned aircraft systems (UAS) and thermal imagery	Drone camera	senseFly	Thermomapper	Not specified	Drone-mounted
2020	Hyun	Remotely piloted aircraft system (Rpas)-based wildlife detection: a review and case studies in maritime Antarctica	Drone camera	FLIR	Vue Pro R	Not specified	Drone-mounted
2020	Smith	Efficacy of aerial forward-looking infrared surveys for detecting Polar Bear maternal dens	Camera	FLIR	Star Safire (models II, III, and HD 380)	Not specified	Aerial (Plane)
2021	Florko	Narwhal (Monodon monoceros) detection by infrared flukeprints from aerial survey imagery	Camera	FLIR	T1020sc	1024 × 768	Aerial (Plane)

Table 4 Literature including the use of thermal imaging for birds between 1968 and 2021 (inclusive).

Year	Author	Title	Device type	Make	Model	Resolution	Deployment type
1982	Best	Aerial thermal infrared census of Canada geese in South Dakota	Line Scanner	Daedalus	Not specified	N/A	Aerial (Plane)
1993	Sidle	Aerial thermal infrared imaging of sandhill cranes on the Platte river, Nebraska	Not specified	Loral	AN/AAD-5	Not specified	Aerial (Plane)
1995	Boonstra	Limitations of far infrared thermal imaging in locating birds	Not specified	Agema	Thermovision 210	Not specified	Ground-based – static
1995	Garner	Use of modern infrared thermography for wildlife population surveys	Scanner	FLIR	2000 A/B	Not specified	Aerial (Plane)
1995	Liechti	Quantification of nocturnal bird migration by moonwatching: comparison with radar and infrared observations	Not specified	Inframetrics	LORIS IRTV-445L	Not specified	Ground-based – static
1996	Benshemesh	Surveying breeding densities of Malleefowl using an airborne thermal scanner	Two: Scanner/ Unspecified	Daedalus/ Inframetrics	1240/60/445	Not specified	Aerial (Plane)
1998	McCafferty	The use of IR thermography to measure the radiative temperature and heat loss of a barn owl (*Tyto alba*)	Camera	AGEMA	AGA 782	Not specified	Not specified
1999	Fortin	Variation in the nocturnal flight behaviour of migratory birds along the northwest coast of the Mediterranean Sea	Not specified	Inframetrics	LORIS 445L	Not specified	Ground-based – static
2001	Zehnder	Nocturnal autumn bird migration at Falsterbo, south Sweden	Camera	Inframetrics	LORIS 445L	Not specified	Ground-based – static
2002	Ovadia	Thermal imaging of House Sparrow nestlings: the effect of begging behavior and nestling rank	Camera	Inframetrics	760	Not specified	Not specified
2003	Galligan	Using a thermographic imager to find nests of grassland birds	Not specified	FLIR	ThermaCAM PM575	Not specified	Ground-based – roaming
2003	Liechti	Nocturnal bird migration in Mauritania – first records	Not specified	Inframetrics	LORIS IRTV-445L	Not specified	Ground-based – static

Year	Author	Title	Device type	Make	Model	Resolution	Deployment type
2005	Lavers	Application of remote thermal imaging and night vision technology to improve endangered wildlife resource management with minimal animal distress and hazard to humans	Camera	Raytheon	PalmIR Pro	320 × 256	Tripod scaled with normal cameras on same target
2006	Gauthreaux	Monitoring bird migration with a fixed-beam radar and a thermal imaging camera	Camera	Raytheon	TI	640 × 482	Tripod mount pointing vertically up
2006	Locke	Evaluation of Portable Infrared Cameras for Detecting Rio Grande Wild Turkeys	Camera	FLIR	ThermaCAMt B-20	320 × 240 pixel	Aerial (Helicopter)
2009	Lazarevic	Improving the efficiency and accuracy of nocturnal bird surveys through equipment selection and partial automation	Imager	FLIR	ThermaCAM P60	320 × 240	Not specified
2010	Conner	Effects of mesopredators on nest survival of shrubnesting songbirds	Not specified	Raytheon	PALM IR 250	Not specified	Vehicle-mounted
2010	O'Neal	Waterfowl on weather radar: applying ground-truth to classify and quantify bird movements	Camera	FLIR	S-60	320 × 240	Tripod Mount
2011	Mills	Capturing clapper rails using thermal imaging technology	Cameras	L-3 Communications	Thermal-Eye 250D; Thermal-Eye X200xp	Not specified	Handheld on airboat
2012	Steen	Automatic Detection of Animals in Mowing Operations Using Thermal Cameras	Camera	FLIR	Not specified	Not specified	Fitted on mowing tractor
2014	Christiansen	Automated detection and recognition of wildlife using thermal cameras	Camera	FLIR	Not specified	320 × 240	Fitted on lift
2015	Chretien	Wildlife multispecies remote sensing using visible and thermal infrared imagery acquired from an unmanned aerial vehicle (UAV)	Uncooled VOx microbolometer	FLIR	Tau640	640 × 480	Fixed to UAV
2015	Cullinan	Classification of birds and bats using flight tracks	Camera	Axsys	FieldPro 5X	Not specified	Fixed on Tripod

Year	Author	Title	Device type	Make	Model	Resolution	Deployment type
2015	Horton	A comparison of traffic estimates of nocturnal flying animals using radar, thermal imaging, and acoustic recording	Camera	FLIR	Guardsman HG-307 Pro	320 × 240	Tripod
2015	Matzner	Two-dimensional thermal video analysis of offshore bird and bat flight	Not specified	Not specified	Not specified	Not specified	Not specified
2016	Austin	If waterbirds are nocturnal are we conserving the right habitats?	Camera	Thermopro	TP8	384 × 288	Tripod-mounted
2017	Dolan	Techniques: Thermal birding	Camera	Pulsar	XD50S	Not specified	Ground-based – roaming
2019	Mitchell	Using infrared thermography to detect night-roosting birds	Camera	InfReC	G120EX	320 × 240	Ground-based – roaming
2019	Scholten	Real-time thermal imagery from an unmanned aerial vehicle can locate ground nests of a grassland songbird at rates similar to traditional methods	Drone camera	FLIR	XT Zenmuse	640 × 512	Drone-mounted
2019	Stephenson	Quantifying thermal-imager effectiveness for detecting bird nests on farms	Camera	FLIR	E8	320 × 240	Ground-based – roaming
2020	Anderson	Deviation Factors in the Mississippi Flyway: Geographic Barriers and Ecological Quality	Security camera	FLIR	SR-19	320(H) × 240(V)	Ground-based – static
2020	Fletcher	Observations on breeding and dispersal by Capercaillie in Strathspey	Scope	Pulsar	Quantum XP50	Not specified	Ground-based – roaming
2020	Graveley	Using a thermal camera to measure heat loss through bird feather coats	Thermal camera	FLIR	SC655	680 × 480	Ground-based – static
2020	Helvey	Duck Nest Detection Through Remote Sensing	Drone camera	Tamarisk	640 Longwave Infrared	Not specified	Drone-mounted
2020	Santangeli	Integrating drone-borne thermal imaging with artificial intelligence to locate bird nests on agricultural land	Drone camera	FLIR/DJI	Tau 2	336 × 256	Drone-mounted
2020	Schedl	Airborne optical sectioning for nesting observation	Drone camera	FLIR	Vue Pro	Not specified	Drone-mounted

Year	Author	Title	Device type	Make	Model	Resolution	Deployment type
2020	Shewring	Using UAV-mounted thermal cameras to detect the presence of nesting nightjar in upland clear-fell: A case study in South Wales, UK	Thermal camera attached to drone	Falcon/FLIR/DJI/ FLIR	8/Tau 2/T600 Inspire 1/ Zenmuse XT V2.0	640 × 512 9 Hz/640 × 512 30 Hz	Drone-mounted
2020	Shirai	The Effectiveness of Visual Scaring Techniques Against Grey Herons, *Ardea cinerea*	Thermal camera	FLIR	Scout TK	Not specified	Not specified
2021	Anderson	Flight directions of songbirds are unaffected by the topography of Lake Erie's southern coastline during fall migration	Camera (Security)	FLIR	SR-19	320 × 240	Ground-based – roaming
2021	Boczkowski	Analysis of the Possibility of Using Unmanned Aerial Vehicles and Thermovision for the Stocktaking of Big Game	Scopes and smartphone attachment and Drone-mounted Camera and Binocular	Pulsar Pulsar HKVision Seek	Thermion XP50 Accolade 2 XP50 LRF Pro Lynx LH15 Thermal Compact Pro	640 × 480 640 × 480 384 × 284 320 × 240	All handheld
2021	Bushaw	Application of unmanned aerial vehicles and thermal imaging cameras to conduct duck brood surveys	Drone Camera	FLIR	Zenmuse XT	640 × 512	Drone-mounted
2021	Corcoran, A. J.	ThruTracker: open-source software for 2-D and 3-D animal video tracking	Drone Camera	FLIR	Zenmuse XT and A65	Not specified	Not specified
2021	Medolago	Use of a portable thermograph as a potential tool to identify nocturnal airport bird risks	Camera	FLIR	T650sc	640 × 480	Ground-based – roaming
2021	Rahman	Performance of unmanned aerial vehicle with thermal imaging, camera trap, and transect survey for monitoring of wildlife	Drone Camera	FLIR	Vue Pro	640 × 360	Drone-mounted
2021	Syposz	Avoidance of different durations, colours and intensities of artificial light by adult seabirds	Camera	FLIR	T640	Not specified	Ground-based – static and Aerial (Helicopter)

Table 5 Literature including the use of thermal imaging for bats between 1968 and 2021 (inclusive).

Year	Author	Title	Device type	Make	Model	Resolution	Deployment type
1991	Kirkwood	Behavioural observations in thermal imaging of the big brown bat, *Eptesicus fuscus*	Not specified	Inframetrics	Imaging radiometer	Not specified	Not specified
1993	Kirkwood	Comparison of two systems for viewing bat behaviour in the dark	infrared imaging radiometer	Inframetrics	525	Not specified	Ground-based – static
1995	Sabol	Technique Using Thermal Infrared-Imaging for Estimating Populations of Gray Bats	Scanning radiometer	Agema	782	Not specified	Ground-based – static
1997	Lancaster	Wing temperature in flying bats measured by infrared thermography	Not specified	Agema	Thermovision 880	Not specified	Ground-based – static
2003	Frank	Advanced infrared detection and image processing for automated bat censusing	Camera	Indigo	Merlin	320 × 256	Ground-based – static
2008	Betke	Thermal imaging reveals significantly smaller Brazilian free-tailed bat colonies than previously estimated	Camera	Merlin	Mid infrared cameras	320 × 240	Tripod-mounted
2008	Horn	Behavioral responses of bats to operating wind turbines	Camera	FLIR	Microbolometer TIR cameras	320 × 240	Tripod-mounted
2009	Ammerman	Census of the endangered Mexican long-nosed bat *Leptonycteris nivalis* in Texas, USA, using thermal imaging	Camera	FLIR	S65	320 × 256	Tripod-mounted
2010	Hristov	Seasonal variation in colony size of Brazilian free-tailed bats at Carlsbad Cavern based on thermal imaging	Camera	FLIR/Indigo Systems	Merlin	320 × 256	Not specified
2010	Reichard	Thermal Windows on Brazilian Free-tailed Bats Facilitate Thermoregulation during Prolonged Flight	Camera	Merlin	Mid TIR camera	Not specified	Not specified
2013	Yang	Study of bat flight behavior by combining thermal image analysis with a LiDAR forest reconstruction	Camera	FLIR	ThermoVision SC8000	1024 X1024	Ground-based static

Year	Author	Title	Device type	Make	Model	Resolution	Deployment type
2014	Cryan	Behavior of bats at wind turbines	Camera	Axis	Q1921-E	384 × 288	Placed looking up towards sky on ground at base of wind turbines
2014	Wu	A thermal infrared video benchmark for visual analysis	Camera	FLIR	SC8000	320 × 240; 1024 × 1024	Tripod, docked in corner of corridors and rooms
2015	Berthinussen	WC1060 Development of a cost-effective method for monitoring the effectiveness of mitigation for bats crossing linear transport infrastructure	Camera	FLIR	H Series	320 × 240	Tripod
2015	Cullinan	Classification of birds and bats using flight tracks	Camera	Axsys	FieldPro 5X	Not specified	Fixed on Tripod
2015	Ganow	Use of thermal imaging to estimate the population sizes of Brazilian free-tailed bat, *Tadarida brasiliensis*, maternity roosts in Oklahoma	Camera	FLIR	SR-19	Not specified	Ground-based – static
2015	Horton	A comparison of traffic estimates of nocturnal flying animals using radar, thermal imaging, and acoustic recording	Camera	FLIR	Guardsman HG-307 Pro	320 × 240	Tripod
2015	Matzner	Two-dimensional thermal video analysis of offshore bird and bat flight	Not specified	Not specified	Not specified	Not specified	Not specified
2016	Diehl	Evaluating the effectiveness of wildlife detection and observation technologies at a solar power tower facility	Camera and Surveillance camera	FLIR/AXIS	SC8343HD/ Q1921-E	Not specified	Ground-based static
2017	Bartonička	Deeply torpid bats can change position without elevation of body temperature	Camera	Guide	M8	Not specified	Not specified
2017	Hayman	Long-term video surveillance and automated analyses reveal arousal patterns in groups of hibernating bats	Surveillance cameras	Axis	Q-1921E	382 × 288	Ground-based – static

Year	Author	Title	Device type	Make	Model	Resolution	Deployment type
2018	Kinzie	Ultrasonic Bat Deterrent Technology	Camera	AXIS	Q1922-E and Q1932_E	Not specified	Ground-based – static
2019	Gilmour	Evaluating methods to deter bats	Thermal camera	Optris	PI640	640 × 480	Ground-based – static
2019	Otálora-Ardila	Thermally-assisted monitoring of bat abundance in an exceptional cave in Brazil's Caatinga drylands	Handheld imager	FLIR	E60	320 × 256	Ground-based – static
2020	Schirmacher	Evaluating the effectiveness of an ultrasonic acoustic deterrent in reducing bat fatalities at wind energy facilities	Thermal camera	AXIS	Q1932-e	640 × 480	Ground-based – static
2020	Simmons	Big brown bats are challenged by acoustically-guided flights through a circular tunnel of hoops	Thermal camera	Photon	320	Not specified	Not specified
2020	Smallwood	Effects of Wind Turbine Curtailment on Bird and Bat Fatalities	Thermal camera	FLIR	T620	Not specified	Not specified
2020	Smallwood	Relating bat passage rates to wind turbine fatalities	Thermal camera	FLIR	T620	640 × 480	Not specified
2020	Smallwood	Dogs detect larger wind energy effects on bats and birds	Not specified	Not specified	Not specified	Not specified	Not specified
2021	Corcoran, A. J.	ThruTracker: open-source software for 2-D and 3-D animal video tracking	Drone Camera	FLIR	Zenmuse XT and A65	Not specified	Not specified
2021	McCarthy	Drone-based thermal remote sensing provides an effective new tool for monitoring the abundance of roosting fruit bats	Drone Camera	FLIR	Zenmuse XT	640 × 520	Drone-mounted

Table 6 Literature including the use of thermal imaging for marsupials between 1968 and 2021 (inclusive).

Year	Author	Title	Device type	Make	Model	Resolution	Deployment type
2000	Dymond	Optimizing the airborne thermal detection of possums	Camera	FLIR	Prism DS/M2000	Not specified	Aerial (Helicopter)
2002	Grierson	The use of aerial digital imagery for kangaroo monitoring	(Line?) Scanner	Redpath Technical Services	Custom device	Not specified	Aerial (Aeroplane)
2016	Gonzalez	Unmanned aerial vehicles (UAVs) and artificial intelligence revolutionizing wildlife monitoring and conservation	Drone camera	FLIR	Tau 2 640	640 × 480	Drone-mounted
2016	Hampton	An assessment of animal welfare for the culling of peri-urban kangaroos	Scope	Guide	518C	Not specified	Not specified
2019	Lethbridge	Estimating kangaroo density by aerial survey: a comparison of thermal cameras with human observers	Thermal camera attached to drone	FLIR	T1K	1024 × 768	Aerial (Helicopter)
2019	Narayan	Using Thermal Imaging to Monitor Body Temperature of Koalas (*Phascolarctos cinereus*) in A Zoo Setting	Thermal camera	FLIR	530	Not specified	Ground-based – handheld
2020	Corcoran	New technologies in the mix: Assessing N-mixture models for abundance estimation using automated detection data from drone surveys	Drone camera	FLIR	Tau 2 640	Not specified	Drone-mounted
2020	Hamilton	Automated Detection of Koalas on Kangaroo Island, South Australia	Drone camera	Not specified	Not specified	640 × 512	Drone-mounted
2020	Vinson	Thermal cameras as a survey method for Australian arboreal mammals: a focus on the greater glider	Handheld imager and thermal camera	FLIR	ThermaCAM S65/T1050sc	320 240/1024 × 768	Ground-based – roaming
2020	Witt	Real-time drone derived thermal imagery outperforms traditional survey methods for an arboreal forest mammal	Drone camera	DJI	Matrice 200 v2	Not specified	Drone-mounted
2021	Baker	Infrared antenna-like structures in mammalian fur	Camera	Leonardo	Merlin	Not specified	Not specified

Year	Author	Title	Device type	Make	Model	Resolution	Deployment type
2021	Corcoran, E.	Evaluating new technology for biodiversity monitoring: Are drone surveys biased?	Drone Camera	FLIR	Tau 2 640	Not specified	Drone-mounted
2021	McGregor	Effectiveness of thermal cameras compared to spotlights for counts of arid zone mammals across a range of ambient temperatures	Scope	FLIR	Scout III 640	640 × 512	Vehicle-based

Table 7 Literature including the use of thermal imaging for insects between 1968 and 2021 (inclusive).

Year	Author	Title	Device type	Make	Model	Resolution	Deployment type
1995	Ono	Unusual thermal defence by a honeybee against mass attack by hornets	Camera	Avio	TVS-8100	Not specified	Not specified
2009	Lavers	Non-destructive high-resolution thermal imaging techniques to evaluate wildlife and delicate biological samples	Camera	FLIR; Echotherm	ThermaCAM E320; Thermal Wave Imaging	320 × 240; 640 × 512	Not specified
2016	Chandra Kumar	Averting Wildlife–Train Interaction Using Thermal Image Processing and Acoustic Techniques	Not specified	Not specified	Not specified	Not specified	Not specified
2019	Roberts	Testing the efficacy of a thermal camera as a search tool for locating wild bumble bee nests	Handheld imager	FLIR	E60	Not specified	Ground-based – roaming

Table 8 Literature including the use of thermal imaging for fish between 1968 and 2021 (inclusive).

Year	Author	Title	Device type	Make	Model	Resolution	Deployment type
1995	Torgerson	Thermal refugia and chinook salmon habitat in Oregon: Applications of airborne thermal videography	Not specified	Agema	Thermovision 800	140 × 140	Aerial (Helicopter)

Resources

This section includes a selection of links to relevant resources to help you find out more and to expand your knowledge on the subject.

Free PDF Guide: ***Top 11 Thermal Imaging Devices for Wildlife***
This PDF guide is freely downloadable. It includes equipment used specifically by wildlife professionals in their work. This includes models suitable for bat survey work.
https://www.wildlifetek.com/top-11-thermal

Guidance Document: ***Thermal Imaging: Bat Survey Guidelines***
This guidance document is freely downloadable as a PDF. Primarily intended for use by ecological consultants surveying for bats, it is also useful for researchers and those working in bat conservation. https://www.bats.org.uk/resources/guidance-for-professionals/thermal-imaging-bat-survey-guidelines

Book: ***Thermal imaging techniques to survey and monitor animals in the wild: a methodology*** **by Havens and Sharp (2015)**
This is an excellent book for anyone interested in gaining a deeper understanding of the principles of thermal imaging for wildlife. Also includes useful information on alternative remote-sensing technologies to consider when screening/scoping a project.
https://www.elsevier.com/books/thermal-imaging-techniques-to-survey-and-monitor-animals-in-the-wild/havens/978-0-12-803384-5

Automation Software: ***ThruTracker***
Developed by Dr Aaron Corcoran and Dr Tyson Hedrick, this computer software programme promises to be an accessible yet powerful solution to the growing volumes of wildlife data requiring analysis. It can automate the detection, tracking and visualisation of wildlife movement from thermal video data. https://sonarjamming.com/thrutracker/

Webinar and Free PDF Guide: ***Thermal Birding***
These excellent resources have been created by West Midlands Bird Ringing Group (WMBRG) and contain valuable information for anyone interested in using thermal imaging for bird survey and monitoring applications.
https://www.westmidlandsringinggroup.co.uk/thermal-birding

Video: ***Tour of the Electromagnetic Spectrum (Infrared Waves)***
This short educational video from NASA gives clear explanations and useful visualisations to help you understand the concepts relating to this part of the electromagnetic spectrum. Well worth a watch! https://science.nasa.gov/ems/07_infraredwaves

Webpage: ***IP Rated Enclosures Explained***
This page explains what IP ratings are and how to read them.
https://www.enclosurecompany.com/ip-ratings-explained.php

Webpage: ***Tattersall Lab***

Dr Glenn Tattersall at Brock University and his diverse team use thermal imaging in their research focusing primarily on thermoregulation and how temperature affects animal lives. Their website has a fascinating collection of blogs and articles on the incredible work they do.

https://tattersalllab.com

Supplier Directory

To help you find relevant services, included below are details of a small number of organisations providing specialist thermal-imaging training, equipment and support.

Wildlifetek
Provides wildlife-specific thermal-imaging training, consulting and support.
UK-based – providing services globally.
https://www.wildlifetek.com

Bat Conservation Trust
Occasionally offers thermal-imaging training for bat survey applications. Proceeds support their incredible bat conservation work.
UK-based.
https://www.bats.org.uk

Vet-IR
Delivers specialist clinical standard physiological imaging, research, training, education and analysis services.
UK-based – providing services globally.
https://www.vet-ir.com

Infrared Training Center
Provides thermal-imaging training, including Level 1, 2 and 3 Thermography certification courses.
Operates globally.
https://www.infraredtraining.com

Wildlife & Countryside Services Ltd
Equipment supplier.
UK-based.
http://wildlifeservices.co.uk

PASS Thermal Ltd
Equipment supplier.
UK-based.
https://www.pass-thermal.co.uk

iRed
Equipment supplier (purchase and hire), calibration and training provider.
UK-based.
https://ired.co.uk

Thermal Vision Research
Equipment supplier (purchase and hire).
UK-based.
https://thermalvisionresearch.co.uk

TJ Focus (Thomas Jacks Limited)
Equipment supplier.
UK-based.
https://tj-focus.co.uk

Veldshop.nl B.V.
Equipment supplier.
Based in the Netherlands.
https://www.veldshop.nl

MoviTHERM
Equipment supplier.
Based in the USA.
https://movitherm.com

References

Allison, N. L., and Destefano, S. (2006) Equipment and techniques for nocturnal wildlife studies. *Wildlife Society Bulletin* 34(4): 1036–1044. DOI: 10.2193/0091-7648(2006)34[1036:EATFNW]2.0.CO;2.

Ammerman, L. K., McDonough, M., Hristov, N. I. and Kunz, T. J. (2009) Census of the endangered Mexican long-nosed bat Leptonycteris nivalis in Texas, USA, using thermal imaging. *Endangered Species Research* 8: 87–92. DOI:10.3354/esr00169.

Amstrup, S. C., York, G., McDonald, T. L., Nielson, R. and Simac, K. (2004) Detecting denning polar bears with forward looking infra-red (FLIR) imagery. *Bioscience* 54(4): 337–344. DOI: 10.1641/0006-3568(2004)054[0337:DDPBWF]2.0.CO;2.

Anand, S., and Radhakrishna, S. (2017) Investigating trends in human–wildlife conflict: is conflict escalation real or imagined? *Journal of Asia-Pacific Biodiversity* 10(2): 154–161. DOI: 10.1016/j.japb.2017.02.003.

Anderson, I. A. (2020) Deviation Factors in the Mississippi Flyway: Geographic Barriers and Ecological Quality. Unpublished MSc thesis, Bowling Green State University.

Anderson, I. A., Gesicki, D. V., and Bingman, V. P. (2021) Flight directions of songbirds are unaffected by the topography of Lake Erie's southern coastline during fall migration. *Journal of Field Ornithology* 92(3): 260–272. DOI: 10.1111/jofo.12374.

Andrade, D. V. (2015) Thermal windows and heat exchange. *Temperature Medical Physiology and Beyond* 2(4): 451. DOI: 10.1080/23328940.2015.1040945.

Anonymous (2010) *Thermography Level 1 Course Manual*. Infrared Training Centre, Danderyd, Sweden.

Anonymous (2016) *Thermography Level 2 Course Manual*. Infrared Training Centre, Danderyd, Sweden.

Arenas, A., Gomez, F., Salas, R., Carrasco, P., Borge, C., Maldonado, A., O'Brien, D., and Martinez-Moreno, F. J. (2002) An evaluation of the application of infrared thermal imaging to the tele-diagnosis of sarcoptic mange in the Spanish ibex (*Capra pyrenaica*). *Veterinary Parasitology* 109: 111–117. DOI: 10.1016/s0304-4017(02)00248-0.

Atuchin, V., Prosekov, A., Vesnina, A., and Kuznetsov, A. (2020) Drone-Assisted Aerial Surveys of Large Animals in Siberian Winter Forests. DOI: 10.21203/rs.3.rs-55278/v1.

Austin, V. I., Ribot, R. F. H., and Bennett, A. T. D. (2016) If waterbirds are nocturnal are we conserving the right habitats? *Emu – Austral Ornithology* 116: 423–427. DOI: 10.1071/MU15106.

Baerwald, E. F., D'Amours, G. H., Klug, B. J., and Barclay, R. M. R. (2008) Barotrauma is a significant cause of bat fatalities at wind turbines. *Current Biology* 18(16): R695-R696. DOI: 10.1016/j.cub.2008.06.029.

Baker, I. M. (2021). Infrared antenna-like structures in mammalian fur. *Royal Society open science* 8(12): 210740. DOI: 10.1098/rsos.210740.

Baldacci, A., Carron, M. J. and Portunato, N. (2005) *Infrared detection of marine mammals*. NATO Undersea Research Centre.

Barber, D. G., Richard, P. R., Hochheim, K. P., and Orr, J. (1991) Calibration of aerial thermal infrared imagery for walrus population assessment. *Arctic* 44(supp. 1): 58–65. DOI: 10.2307/40510982.

Barber, D. G., Richard, P. R., and Hochheim, K. P. (1989) Thermal remote sensing for walrus population assessment in the Canadian Arctic. *International Geoscience and Remote Sensing Symposium and Twelfth Canadian Symposium on Remote Sensing*, Vancouver, BC, Canada 10–14 July.

Barbieri, M. M., Mclellan, W. A., Wells, R. S., Blum, J. E., Hofmann, S., Gannon, J., and Pabst, D. A. (2010) Using infrared thermography to assess seasonal trends in dorsal fin surface temperatures of free-swimming bottlenose dolphins (*Tursiops truncatus*) in Sarasota Bay, Florida. *Marine Mammal Science* 26: 53–66. DOI: 10.1111/j.1748-7692.2009.00319.x.

Barclay, R. M. R., and Harder, L. D. (2003) Life histories of bats: Life in the slow lane. In Kunz, T. H., and Fenton, B. M. (Eds) *Bat Ecology*. University of Chicago Press: Chicago. Pp 209–256.

Bartonička, T., Bandouchova, H., Berková, H., Blažek, J., Lučan, R., Horáček, I., ... and Zukal, J. (2017) Deeply torpid bats can change position without elevation of body temperature. *Journal of Thermal Biology* 63: 119–123. DOI: 10.1016/j.jtherbio.2016.12.005.

BBC (2015) Winterwatch: Blakeney Point seals filmed at night. *BBC*. Accessed at https://www.bbc.co.uk/news/av/uk-england-norfolk-30862195.

BBC (2016) Planet Earth II: Cities. *BBC*. Accessed at https://www.bbc.co.uk/programmes/m000nx12.

BBC (2019) Inside the Bat Cave. *BBC*. Accessed at https://www.bbc.co.uk/programmes/m000nx12.

Beaver, J. T., Baldwin, R. W., Messinger, M., Newbolt, C. H., Ditchkoff, S. S., and Silman, M. R. (2020) Evaluating the Use of Drones Equipped with Thermal Sensors as an Effective Method for Estimating Wildlife. *Wildlife Society Bulletin* 44(2): 434–443. DOI: 10.1002/wsb.1090.

Beaver, J. T., Harper, C. A., Kissell Jr., R. E., Muller, L., Basinger, P. S., Goode, M. J., Van Manen, F. T., Winton, W., and Kennedy, M. L. (2014) Aerial vertical-looking infrared imagery to evaluate bias of distance sampling techniques for white-tailed deer. *Wildlife Society Bulletin* 38(2): 419–427. DOI: 10.1002/wsb.410.

Bedson, C. P., Thomas, L., Wheeler, P. M., Reid, N., Harris, W. E., Lloyd, H., Mallon, D., and Preziosi, R. (2021) Estimating density of mountain hares using distance sampling: a comparison of daylight visual surveys, night-time thermal imaging and camera traps. *Wildlife Biology* 2021(3): wlb-00802. DOI: 10.2981/wlb.00802.

Belant, J. L., and Seamans, T. W. (2000) Comparison of 3 devices to observe white-tailed deer at night. *Wildlife Society Bulletin* 28(1): 154–158. DOI: 10.2307/4617297.

Benshemesh, J. S., and Emison, W. B. (1996) Surveying breeding densities of Malleefowl using an airborne thermal scanner. *Wildlife Research* 23(2): 121–142. DOI: 10.1071/WR9960121.

Bernatas, S., and Nelson, L. (2004) Sightability model for California bighorn sheep in canyonlands using forward-looking infrared (FLIR). *Wildlife Society Bulletin* 32(3): 638–647. DOI: 10.2193/0091-7648(2004)032[0638:SMFCBS]2.0.CO;2.

Berthinussen, A., and Altringham, J. (2015) *WC1060 Development of a cost-effective method for monitoring the effectiveness of mitigation for bats crossing linear transport infrastructure*. UK: Department for Environmental Food and Rural Affairs.

Best, R. G., Hause, D., Wehde, M., and Fowler, R. (1982) Aerial thermal infrared census of Canada geese in South Dakota. *Photogrammetric Engineering and Remote Sensing* 48(12): 1869–1877.

Betke, M., Hedrick, T., and Theriault, D. (2018) Multi-Camera Videography Methods for Aeroecology in Chilson, P., Frick, W., Kelly, J., and Liechti, F. (eds) *Aeroecology*. Springer, Cham. Pp. 239–257. DOI: 10.1007/978-3-319-68576-2_10.

Betke, M., Hirsh, D. E., Makris, N. C., McCracken, G. F., Procopio, M., Hristov, N. I., Teng, S., Bacchi, A., Reichard, J. D., Horn, J. W., Crampton, S., Cleveland, C. J., and Kunz, T. H. (2008) Thermal imaging reveals significantly smaller Brazilian free-tailed bat colonies than previously estimated. *Journal of Mammalogy* 89(1): 18–24. DOI: 10.1644/07-MAMM-A-011.1.

Biondi, F., Dornbusch, P. T., Sampaio, M., and Montiani-Ferreira, F. (2013) Infrared ocular thermography in dogs with and without keratoconjunctivitis sicca. *Veterinary Ophthalmology* 18(1): 28–34. DOI: 10.1111/vop.12086.

Boczkowski, A., Kuboszek, A., Mańka, A., Dutkiewicz, K., Gramatyka, W., Leśniak, A., Spyrka, M., and Stawinoga, E. (2021) Analysis of the Possibility of Using Unmanned Aerial Vehicles and Thermovision for the Stocktaking of Big Game. *Multidisciplinary Aspects of Production Engineering* 4(1): 317–329. DOI: 10.2478/mape-2021-0029.

Boonstra, R., Krebs, C. J., Boutin, S., and Eadie, J. M. (1994) Finding mammals using far-infrared thermal imaging. *Journal of Mammalogy* 75(4): 1063–1068. DOI: 10.2307/1382490.

Boonstra, R., Eadie, J. M., Krebs, C. J., and Boutin, S. (1995) Limitations of far infrared thermal imaging in locating birds. *Journal of Field Ornithology* 66(2): 192–198.

Bossomaier, T. R. (2012) *Introduction to the senses: from biology to computer science*. Cambridge University Press.

Bowen, C., Reeve, N., Pettinger, T., and Gurnell, J. (2020) An evaluation of thermal infrared cameras for surveying hedgehogs in parkland habitats. *Mammalia* 84(4): 354–356. DOI: 10.1515/mammalia-2019-0100.

Brawata, R. L., Raupach, T. H., and Neeman, T. (2013) Techniques for monitoring carnivore behavior using automatic thermal video. *Wildlife Society Bulletin* 37(4): 862–871. DOI: 10.1002/wsb.315.

Brooks, J. W. (1972) Infra-Red Scanning for Polar Bear. *Bears: Their Biology and Management* 2: 138–141. DOI: 10.2307/3872577.

Buonaccorsi, J. and Staudenmayer, J. (2009) Statistical methods to correct for observation error in a density-independent population model. *Ecological Monographs* 79(2): 299–324. DOI: 10.1890/07-1299.1.

Burger, A. L., Fennessy, J., Fennessy, S., and Dierkes, P. W. (2020) Nightly selection of resting sites and group behavior reveal antipredator strategies in giraffe. *Ecology and Evolution* 10(6), 2917–2927. DOI: 10.1002/ece3.6106.

Burke, C., Rashman, M. F., Longmore, S. N., McAree, O., Glover-Kapfer, P., Ancrenaz, M., and Wich, S. A. (2019) Successful observation of orangutans in the wild with thermal-equipped drones. *Journal of Unmanned Vehicle Systems* 99: 1–25. DOI: 10.1139/juvs-2018-0035.

Burke, C., Rashman, M. F., Wich, S., Symons, A., Theron, C., and Longmore, S. (2019) Optimizing observing strategies for monitoring animals using drone-mounted thermal infrared cameras. *International Journal of Remote Sensing* 40(2): 439–467. DOI: 10.1080/01431161.2018.1558372.

Burkhardt E., Kindermann L., Zitterbart D., Boebel O. (2012) Detection and Tracking of Whales Using a Shipborne, 360° Thermal-Imaging System. In: Popper, A. N., Hawkins, A. (eds) *The Effects of Noise on Aquatic Life*. Advances in Experimental Medicine and Biology, vol. 730. Springer, New York, NY. https://doi.org/10.1007/978-1-4419-7311-5_66.

Burn, D. M., Udevitz, M. S., Speckman, S. G., and Benter, R. B. (2009) An improved procedure for detection and enumeration of walrus signatures in airborne thermal imagery. *International Journal of Applied Earth Observation and Geoinformation* 11(5): 324–333. DOI: 10.1016/j.jag.2009.05.004.

Burn, D. M., Webber, M. A., and Udevitz, M. S. (2006) Application of Airborne Thermal Imagery to Surveys of Pacific Walrus. *Wildlife Society Bulletin* 34, 51–58. DOI: 10.2193/0091-7648(2006)34[51:AOATIT]2.0.CO;2.

Bushaw, J. D., Ringelman, K. M., and Rohwer, F. C. (2019) Applications of unmanned aerial vehicles to survey mesocarnivores. *Drones* 3(1): 28. DOI:10.3390/drones3010028.

Butler, D. A., Ballard, W. B., Haskell, S. P., and Wallace, M. C. (2006) Limitations of thermal infrared imaging for locating neonatal deer in semiarid shrub communities. *Wildlife Society Bulletin* 34, 1458–1462. DOI: 10.2193/0091-7648(2006)34[1458:LOTIIF]2.0.CO;2.

Carr, N. L., Rodgers, A. R., Kingston, S. R., Hettinga, P. N., Thompson, L. M., Renton, J. L., and Watson, P. J. (2010) Comparative woodland caribou population surveys in Slate Islands Provincial Park, Ontario. *Rangifer* 32(Special Issue 20): 205–217. DOI: 10.7557/2.32.2.2270.

Cena, K., and Clark, J. A. (1973) Thermographic measurements of the surface temperatures of animals. *Journal of Mammalogy* 54(4): 1003–1007. DOI: 10.2307/1379105.

Chabot, D. and Bird, D. M. (2015) Wildlife research and management methods in the 21st century: Where do unmanned aircraft fit in? *Journal of Unmanned Vehicle Systems* 3: 137–155. DOI: 10.1139/juvs-2015-0021.

Chandra Kumar, J., and Selvakumar, A. (2016) Averting Wildlife–Train Interaction Using Thermal Image Processing and Acoustic Techniques. *Journal of Biodiversity Management & Forestry* 5(3): 11–13. DOI: 10.4172/2327-4417.1000165.

Chapman, L. A., and White, P. C. (2020) Anti-poaching strategies employed by private rhino owners in South Africa. *Pachyderm* 61: 179–183.

Chrétien, L. P., Théau, J., and Ménard, P. (2015) Wildlife multispecies remote sensing using visible and thermal infrared imagery acquired from an unmanned aerial vehicle (UAV). *International Archives of the Photogrammetry, Remote Sensing & Spatial Information Sciences* 40(XL-1/W4): 241–248. DOI: 10.5194/isprsarchives-XL-1-W4-241-2015.

Chrétien, L. P., Théau, J., and Ménard, P. (2016) Visible and thermal infrared remote sensing for the detection of white-tailed deer using an unmanned aerial system. *Wildlife Society Bulletin* 40(1): 181–191. DOI: 10.1002/wsb.629.

Christiansen, P., Steen, K. A., Jørgensen, R. N., and Karstoft, H. (2014) Automated detection and recognition of wildlife using thermal cameras. *Sensors* 14(8): 13778–13793. DOI: 10.3390/s140813778.

Colchester Zoo (2021) Colchester Zoo's elephants take part in HEAT project! *Colchester Zoo*. Accessed at https://www.colchester-zoo.com/2021/02/colchester-zoos-elephants-take-part-in-heat-project/ (24 May 2022).

Collins, J. (2016) *Bat Surveys for Professional Ecologists: Good Practice Guidelines* (3rd edn). The Bat Conservation Trust: London.

Conn, P. B., Ver Hoef, J. M., McClintock, B. T., Moreland, E. E., London, J. M., Cameron, M. F., Dahle, S. P., and Boveng, P. L. (2014) Estimating multispecies abundance using automated detection systems: ice-associated seals in the Bering Sea. *Methods in Ecology and Evolution* 5(12): 1280–1293. DOI: 10.1111/2041-210X.12127.

Conner, L. M., Rutledge, J. C., and Smith, L. L. (2010) Effects of mesopredators on nest survival of shrubnesting songbirds. *The Journal of Wildlife Management* 74(1): 73–80. DOI: 10.2193/2008-406.

Conno, C. D., Nardone, V., Ancillotto, L., De Bonis, S., Guida, M., Jorge, I., Scarpa, U., and Russo, D. (2018) Testing the performance of bats as indicators of riverine ecosystem quality. *Ecological Indicators* 95(1): 741–750. DOI: 10.1016/j.ecolind.2018.08.018.

Corcoran, A. J., Schirmacher, M. R., Black, E., and Hedrick, T. L. (2021) ThruTracker: open-source software for 2-D and 3-D animal video tracking. *bioRxiv*. DOI: 10.1101/2021.05.12.443854.

Corcoran, E., Denman, S., and Hamilton, G. (2021) Evaluating new technology for biodiversity monitoring: Are drone surveys biased? *Ecology and Evolution* 11(11): 6649–6656. DOI: 10.1002/ece3.7518.

Corcoran, E., Denman, S., and Hamilton, G. (2020) New technologies in the mix: Assessing N-mixture models for abundance estimation using automated detection data from drone surveys. *Ecology and Evolution* 10(15): 8176–8185. DOI: 10.1002/ece3.6522.

Cox, T. E., Matthews, R., Halverson, G., and Morris, S. (2021) Hot stuff in the bushes: Thermal imagers and the detection of burrows in vegetated sites. *Ecology and Evolution* 11(11): 6406–6414. DOI: 10.1002/ece3.7491.

Croft, S., Franzetti, B., Gill, R., and Massei, G. (2020) Too many wild boar? Modelling fertility control and culling to reduce wild boar numbers in isolated populations. *PloS one* 15(9), e0238429. DOI: 10.1371/journal.pone.0238429.

Croon, G. W., McCullough, D. R., Olson, C. E., and Queal, L. M. (1968) Infrared scanning techniques for big game censusing. *Journal of Wildlife Management* 32(4): 752–759.

Cryan, P. M., Gorresen, P. M., Hein, C. D., Schirmacher, M. R., Diehl, R. H., Huso, M. M., Hayman, D. T. S., Fricker, P. D., Bonaccorso, F. J., Johnson, D. H., Heist, K., and Dalton, D. C. (2014) Behavior of bats at wind turbines. *Proceedings of the National Academy of Sciences* 111(42): 15126–15131. DOI: 10.1073/pnas.1406672111.

Cui, Y., Gong, H., Wang, Y., Li, D., and Bai, H. (2018) A thermally insulating textile inspired by polar bear hair. *Advanced Materials* 30(14): 1706807. DOI: 10.1002/adma.201706807.

Cullinan, V. I., Matzner, S., and Duberstein, C. A. (2015) Classification of birds and bats using flight tracks. *Ecological Informatics* 27: 55–63. DOI: 10.1016/j.ecoinf.2015.03.004.

Cuthill, I. C. (2019) Camouflage. *Journal of Zoology* 308 (2): 75–92. DOI:10.1111/jzo.12682.

Cuyler, L. C., Wiulsrod, R., and Ortisland, N. A. (1992) Thermal infrared radiation from free living whales. *Marine Mammal Science* 8: 120–134. DOI: 10.1111/J.1748-7692.1992.TB00371.X.

da Costa, A. L. M., Rassy, F. B., and da Cruz, J. B. (2020) Diagnostic applications of Infrared Thermography in Captive Brazilian Canids and Felids. *Archives of Veterinary Science* 25(2). DOI: 10.5380/avs.v25i2.71088.

Dahlen, B., and Traeholt, C. (2018) Successful aerial survey using thermal camera to detect wild orangutans in a fragmented landscape. *Journal of Indonesian Natural History* 6(2): 12–23.

D'Angelo, G. J., D'Angelo, J. G., Gallagher, G. R., Osborn, D. A., Miller, K. V., and Warren, R. J. (2010) Evaluation of Wildlife Warning Reflectors for Altering White-Tailed Deer Behavior Along Roadways. *Wildlife Society Bulletin* 34(4): 1175–1183. DOI: 10.2193/0091-7648(2006)34[1175:EOWWRF]2.0.CO;2.

Dangerfield, A. (2020) Progress report – February 2020 – Thermal imaging for human–wildlife conflict. *Arribada Initiative*. Accessed at https://arribada.org/2020/02/17/progress-report-feburart-2020-thermal-imaging-for-human-wildlife-conflict/ (24 May 2022).

de Queiroz, J. P. A. F., Souza-Junior, J. B. F., Oliviera, V. R. M., Costa, L. L. M., and Oliviera, M. F. (2020) How does Spix's yellow-toothed cavy (*Galea spixii Wagler, 1831*) face the thermal challenges of the Brazilian tropical dry forest? *Journal of Thermal Biology* 88: 102525. DOI: 10.1016/j.jtherbio.2020.102525.

Department for Environment Food and Rural Affairs. (2013) *Monitoring the humaneness of badger population reduction by controlled shooting.*

Dezecache, G., Wilke, C., Richi, N., Neumann, C., and Zuberbuhler, K. (2017) Skin temperature and reproductive condition in wild female chimpanzees. *PeerJ* 5: e4116. DOI: 10.7717/peerj.4116.

Diehl, R. H., Valdez, E. W., Preston, T. M., Wellik, M. J., and Cryan, P. M. (2016) Evaluating the effectiveness of wildlife detection and observation technologies at a solar power tower facility. *PloS one* 11(7): e0158115. DOI: 10.1371/journal.pone.0158115.

Ditchkoff, S. S., Raglin, J. B., Smith, J. M. and Collier, B. A. (2005) From the Field: capture of white-tailed deer fawns using thermal imaging technology. *Wildlife Society Bulletin* 33(3): 1164–1168.

Dolan, B., and Brewood Ringers (2017) Techniques: Thermal birding. *LifeCycle Magazine* 5: 16–18.

Domino, M., Borowska, M., Kozłowska, N., Zdrojkowski, Ł., Jasiński, T., Smyth, G., and Maśko, M. (2022) Advances in Thermal Image Analysis for the Detection of Pregnancy in Horses Using Infrared Thermography. *Sensors* 22(1): 191. DOI: 10.3390/s22010191.

Drake, D., Aquila, C., and Huntington, G. (2005) Counting a suburban deer population using forward looking infrared radar and road counts. *Wildlife Society Bulletin* 33(2): 656–661. DOI: 10.2193/0091-7648(2005)33%5B656:CASDPU%5D2.0.CO;2.

Duffy, R., St John, F. A., Büscher, B., and Brockington, D. (2016) Toward a new understanding of the links between poverty and illegal wildlife hunting. *Conservation Biology* 30(1), 14–22. DOI: 10.1111/cobi.12622.

Dunbar, M., Johnson, S. R., Rhyan, J. C., McCollum, M. (2009) Use of infrared thermography to detect thermographic change in mule deer (*Odocoileus hemionus*) experimentally infected with foot-and-mouth disease. *Journal of Zoo and Wildlife Medicine* 40(2): 296–301. DOI: 10.1638/2008-0087.1.

Dunbar, M. R., and MacCarthy, K. A. (2006) Use of infrared thermography to detect signs of rabies infection in raccoons (Procyon lotor). *Journal of Zoo and Wildlife Medicine* 37(4): 518–523. DOI: 10.1638/06-039.1.

Dunn, W. C., Donnelly, J. P., and Krausmann, W. J. (2002) Using thermal infrared sensing to count elk in the south-western United States. *Wildlife Society Bulletin* 30: 963–967. DOI: 10.2307/3784254.

Dymond, J. R., Trotter, C. M., Shepherd, J. D., and Wilde, H. (2000) Optimizing the airborne thermal detection of possums. *International Journal of Remote Sensing* 21: 3315–3326. DOI: 10.1080/014311600750019921.

Eddy, A. L., Van Hoogmoed, L. M., and Snyder, J. R. (2001) The Role of Thermography in the Management of Equine Lameness. *The Veterinary Journal* 162(3): 172–181. DOI: 10.1053/tvjl.2001.0618.

Edmunds, M. (2004) Crypsis. In: *Encyclopedia of Entomology*. Springer, Dordrecht. DOI: 10.1007/0-306-48380-7_1087.

Faber Johannesen, L. (2018) The efficiency of thermal imaging for wildlife applications. Bachelor's Thesis. University of Copenhagen.

Faber Johannesen, L. (2021) Wildlife vs. infrastructure: How could we reduce conflicts? *Wildlifetek KFW Scientific & Creative*. Accessed at https://www.wildlifetek.com/blog/wildlifevsinfrastructure (26 May 2022).

Fawcett Williams, K. (2021) *Thermal Imaging: Bat Survey Guidelines*. In association with the Bat Conservation Trust.

Felton, M., Gurton, K. P., Pezzaniti, J. L., Chenault, D. B., and Roth, L. E. (2010) Measured comparison of the crossover periods for mid- and long-wave IR (MWIR and LWIR) polarimetric and conventional thermal imagery. *Optics Express* 18(15): 15704–15713. DOI: 10.1364/OE.18.015704.

Filipovs, J., Berg, A., Ahlberg, J., Vecvanags, A., Brauns, A., and Jakovels, D. (2021) UAV areal imagery-based wild animal detection for sustainable wildlife management. In *EGU General Assembly Conference Abstracts* (pp. EGU21–14137).

Fletcher, K., and Baines, D. (2020) Observations on breeding and dispersal by Capercaillie in Strathspey. *Scottish Birds* 27(34): 34.

FLIR (2020) FLIR Security Solutions Protect Munu, an Endangered South-Western Black Rhino. *Teledyne FLIR LLC*. Accessed at https://www.flir.co.uk/discover/security/flir-security-solutions-protect-munu-an-endangered-south-western-black-rhino/ (15 May 2022).

FLIR (2021, 1 November) Cooled or Uncooled? *Teledyne FLIR LLC*. Accessed at https://www.flir.co.uk/discover/rd-science/cooled-or-uncooled/ (1 May 2022).

FLIR (n.d.) Thermal Imaging Cameras Used For Deer Management. *Teledyne FLIR LLC*. Accessed at https://www.flir.co.uk/discover/ots/outdoor/thermal-imaging-cameras-used-for-deer-management/ (15 May 2022).

Flodell, A., and Christensson, C. (2016) *Wildlife surveillance using a uav and thermal imagery*. Unpublished MSc Thesis, Linköping University, Sweden.

Florko, K. R., Carlyle, C. G., Young, B. G., Yurkowski, D. J., Michel, C., and Ferguson, S. H. (2021) Narwhal (*Monodon monoceros*) detection by infrared flukeprints from aerial survey imagery. *Ecosphere* 12(8): e03698. DOI: 10.1002/ecs2.3698.

Focardi, S., De Marinis, A. M., Rizzotto, M., and Pucci, A. (2001) Comparative evaluation of thermal infrared imaging and spotlighting to survey wildlife. *Wildlife Society Bulletin* 29(1): 133–139. DOI: 10.2307/3783989.

Fortin, D., Liechti, F., and Bruderer, B. (1999) Variation in the nocturnal flight behaviour of migratory birds along the northwest coast of the Mediterranean Sea. *Ibis* 141: 480–488. DOI: 10.1111/J.1474-919X.1999. TB04417.X.

Frank, J. D., Kunz, T. H., Horn, J., Cleveland, C., and Petronio, S. (2003) Advanced infrared detection and image processing for automated bat censusing. *Infrared technology and applications XXIX*. Proceedings of SPIE 5074: 261–271. DOI: 10.1117/12.499422.

Franke, U., Goll, B., Hohmann, U., and Heurich, M. (2012) Aerial ungulate surveys with a combination of infrared and high-resolution natural colour images. *Animal Biodiversity and Conservation* 35(2): 285–293. DOI: 10.32800/abc.2012.35.0285.

Franzetti, B., Ronchi, F., Marini, F., Scacco, M., Calmanti, R., Calabrese, A., Paola, A., Paolo, M., and Focardi, S. (2012) Nocturnal line transect sampling of wild boar (*Sus scrofa*) in a Mediterranean forest: long-term comparison with capture-mark-resight population estimates. *European Journal of Wildlife Research* 58(2): 385–402. DOI: 10.1007/s10344-011-0587-x.

Galligan, E. W., Bakken, G. S., and Lima, S. L. (2003) Using a thermographic imager to find nests of grassland birds. *Wildlife Society Bulletin* 31(3): 865–869. DOI: 10.2307/3784611.

Ganow, K. B., Caire, W., and Matlock, R. S. (2015) Use of thermal imaging to estimate the population sizes of Brazilian free-tailed bat, Tadarida brasiliensis, maternity roosts in Oklahoma. *The Southwestern Naturalist* 60(1): 90–96. DOI: 10.1894/SWNAT-D-14-00010R1.1.

Garner, D. L., Underwood, H. B., and Porter, W. F. (1995) Use of modern infrared thermography for wildlife population surveys. *Environmental Management* 19(2): 233–238. DOI: 10.1007/BF02471993.

Gao, B., Hedlund, J., Reynolds, D. R., Zhai, B., Hu, G., and Chapman, J. W. (2020) The 'migratory connectivity' concept, and its applicability to insect migrants. *Movement Ecology* 8; 42. DOI: doi.org/10.1186/s40462-020-00235-5.

Gauthreaux, S. A. Jr, and Livingston, J. W. (2006) Monitoring bird migration with a fixed-beam radar and a thermal imaging camera. *Journal of Field Ornithology* 77(3): 319–328 DOI: 10.1111/j.1557-9263.2006.00060.x.

Giannetto, C., Di Pietro, S., Falcone, E., Pennisi, M., Giudice, E., Piccione, G., and Acri, G. (2021) Thermographic ocular temperature correlated with rectal temperature in cats. *Journal of Thermal Biology* 102: 103104. DOI: 10.1016/j.jtherbio.2021.103104.

Gill, R. M. A., Thomas, M. L. and Stocker, D. (1997) The use of portable thermal imaging for estimating deer population density in forest habitats. *Journal of Applied Ecology* 34: 1273–1286. DOI: 10.2307/2405237.

Gilmour, L. R. (2019) Evaluating methods to deter bats. Unpublished PhD Thesis, University of Bristol.

Gordon, M., and Weber, M. (2021) Global energy demand to grow 47% by 2050, with oil still top source: US EIA. *S & P Commodity Insights*. Accessed at https://www.spglobal.com/commodityinsights/en/market-insights/latest-news/oil/100621-global-energy-demand-to-grow-47-by-2050-with-oil-still-top-source-us-eia (24 May 2022).

Gonzalez, L. F., Montes, G. A., Puig, E., Johnson, S., Mengersen, K., and Gaston, K. J. (2016) Unmanned aerial vehicles (UAVs) and artificial intelligence revolutionizing wildlife monitoring and conservation. *Sensors* 16(1): 97. DOI: 10.3390/s16010097.

Goodale, M. W., and Milman, A. (2016) Cumulative adverse effects of offshore wind energy development on wildlife. *Journal of Environmental Planning and Management* 59(1): 1–21. DOI: 10.1080/09640568.2014.973483.

Gooday, O. J., Key, N., Goldstien, S., and Zawar-Reza, P. (2018) An assessment of thermal-image acquisition with an unmanned aerial vehicle (UAV) for direct counts of coastal marine mammals ashore. *Journal of Unmanned Vehicle Systems* 6(2): 100–108. DOI: 10.1139/juvs-2016-0029.

Goodenough, A. E., Carpenter, W. S., MacTavish, L., Theron, C., Delbridge, M., and Hart, A. G. (2018) Identification of African Antelope Species: Using Thermographic Videos to Test the Efficacy of Real-Time Thermography. *African Journal of Ecology* 56(4): 898–907. DOI: 10.1111/aje.12513.

Graber, J (2011) Land-based infrared imagery for marine mammal detection. Masters Thesis. University of Washington.

Graber, J., Thomson, J., Polagye, B., and Jessup, A. (2011) *Land-based infrared imagery for marine mammal detection*, Proceedings SPIE 8156, Remote Sensing and Modeling of Ecosystems for Sustainability VIII, 81560B. DOI: 10.1117/12.892787.

Graveley, J. M., Burgio, K. R., and Rubega, M. (2020) Using a thermal camera to measure heat loss through bird feather coats. *Journal of Visualized Experiments* 160: e60981. DOI: 10.3791/60981.

Graves, H. B., Bellis, E. D., and Knuth, W. M. (1972) Censusing white-tailed deer by airborne thermal infrared imagery. *Journal of Wildlife Management* 36(3): 875–884. DOI: 10.2307/3799443.

Gregory, A., Spence, E., Beier, P., and Garding, E. (2021) Toward Best Management Practices for Ecological Corridors. *Land*, 10: 140. DOI: 10.3390/land10020140.

Greggor, A. L., Blumstein, D. T., Wong, B. B., and Berger-Tal, O. (2019) Using animal behavior in conservation management: a series of systematic reviews and maps. *Environmental Evidence* 8(1): 23. DOI: 10.1186/s13750-019-0164-4.

Grierson, I. T., and Gammon, J. A. (2002) The use of aerial digital imagery for kangaroo monitoring. *Geocarto International* 17(2): 43–49. DOI: 10.1080/10106040208542234.

Gurnell, J., Reeve, N., and Bowen, C. (2021) Surveys of Hedgehogs in The Regent's Park, London. *The Royal Parks*. Accessed at https://www.royalparks.org.uk/__data/assets/pdf_file/0005/129479/The-Regents-Park-Hedgehog-Research-Report-2014-2020.pdf (15 May 2022).

Hambrecht, L., Brown, R. P., Piel, A. K., and Wich, S. A. (2019) Detecting 'poachers' with drones: Factors influencing the probability of detection with TIR and RGB imaging in miombo woodlands, Tanzania. *Biological Conservation* 233: 109–117. DOI: 10.1016/j.biocon.2019.02.017.

Hamilton, G., Corcoran, E., Winsen, M., and Denman, S. (2020) *Automated Detection of Koalas on Kangaroo Island*. Accessed at https://eprints.qut.edu.au/209098/ (17 May 2022).

Hampton, J. O., and Forsyth, D. M. (2016) An assessment of animal welfare for the culling of peri-urban kangaroos. *Wildlife Research* 43(3): 261–266. DOI: 10.1071/WR16023.

Haroldson, B. S., Wiggers, E. P., Beringer, J., Hansen, L. P., and McAninch, J. B. (2003) Evaluation of aerial thermal imaging for detecting white-tailed deer in a deciduous forest environment. *Wildlife Society Bulletin* 31: 1188–1197. DOI: 10.2307/3784466.

Haroldson, B. S. (1999) Evaluation of thermal infrared imaging for detection of white-tailed deer. Unpublished MSc thesis, University of Missouri.

Harrap, M. J. M., de Ibarra, N. H., Whitney, H. M., and Rands, S. A. (2018) Reporting of thermography parameters in biology: a systematic review of thermal imaging literature. *Royal Society Open Science* 5: 181281. DOI: 10.1098/rsos.181281.

Hart, A. G., Rolfe, R. N., Dandy, S., Stubbs, H., MacTavish, D., MacTavish, L., and Goodenough, A. E. (2015) Can handheld thermal imaging technology improve detection of poachers in African bushveldt? *PLoS One* 10(6): e0131584. DOI: 10.1371/journal.pone.0131584.

Havens, K. J., and Sharp, E. J. (1995) The use of thermal imagery in the aerial survey of panthers (and other animals) in the Florida Panther National Wildlife Refuge and the Big Cypress National Preserve. Final Report to U.S. Fish and Wildlife Service (Naples, Florida).

Havens, K. J., and Sharp, E. J. (1998) Using thermal imagery in the aerial survey of animals. *Wildlife Society Bulletin* 26: 17–23.

Havens, K. J., and Sharp, E. (2015) *Thermal imaging techniques to survey and monitor animals in the wild: a methodology*. Academic Press.

Hayman, D. T., Cryan, P. M., Fricker, P. D., and Dannemiller, N. G. (2017) Long-term video surveillance and automated analyses reveal arousal patterns in groups of hibernating bats. *Methods in Ecology and Evolution* 8(12): 1813–1821. DOI: 10.1111/2041-210X.12823.

Heintz, M. R., Fuller, G., and Allard, S. (2019) Exploratory investigation of infrared thermography for measuring gorilla emotional responses to interactions with familiar humans. *Animals* 9(9): 604. DOI: 10.3390/ani9090604.

Helvey, M., Ryckman, M., Ellis-Felege, S., Van Aardt, J., and Salvagio, C. (2020) Duck Nest Detection Through Remote Sensing. In *IGARSS 2020 – 2020 IEEE International Geoscience and Remote Sensing Symposium*: 6321–6324. IEEE. DOI: 10.1109/IGARSS39084.2020.9323500.

Hemami, M. R., Watkinson, A. R., Gill, R. M. A., and Dolman, P. M. (2007) Estimating abundance of introduced Chinese muntjac *Muntiacus reevesi* and native roe deer *Capreolus capreolus* using portable thermal imaging equipment. *Mammal Review* 37(3): 246–254. DOI: 10.1111/j.1365-2907.2007.00110.x.

Hilsberg-Merz, S. (2008) Infrared thermography in zoo and wild animals. *Zoo and wild animal medicine current therapy* 6: 20–33.

Hodnett, E. (2005) Thermal imaging applications in urban deer control in (Nolte, D. L., Fagerstone, K. A. Eds) *Wildlife Damage Management Conferences – Proceedings*. 106.

Hohmann, U., Kronenberg, M., Scherschlicht, M., and Schönfeld, F. (2021) The possibilities and limitations of thermal imaging to detect wild boar (*Sus scrofa*) carcasses as a strategy for managing African Swine Fever (ASF) outbreaks. *Berliner und Münchener Tierärztliche Wochenschrift* 134: 1–14. DOI: 10.2376/1439-0299-2020-46.

Horn, J. W., Arnett, E. B., and Kunz, T. H. (2008) Behavioral responses of bats to operating wind turbines. *The Journal of Wildlife Management* 72(1): 123–132. DOI: 10.2193/2006-465.

Horns, J. J. and Şekercioğlu, C. (2018) Conservation of migratory species. *Current Biology* 28(17): R980-R983. DOI: 10.1016/j.cub.2018.06.032.

Horton, K. G., Shriver, W. G., and Buler, J. J. (2015) A comparison of traffic estimates of nocturnal flying animals using radar, thermal imaging, and acoustic recording. *Ecological Applications* 25(2): 390–401. DOI: 10.1890/14-0279.1.

Hristov, N. L., Betke, M., Theriault, D. E. H., Bagchi, A. and Kunz, T. H. (2010) Seasonal variation in colony size of Brazilian free-tailed bats at Carlsbad Cavern based on thermal imaging. *Journal of Mammalogy* 91(1): 183–192. DOI: 10.1644/08-MAMM-A-391R.1.

Hyun, C. U., Park, M., and Lee, W. Y. (2020) Remotely piloted aircraft system (Rpas)-based wildlife detection: a review and case studies in maritime Antarctica. *Animals* 10(12): 2387. DOI: 10.3390/ani10122387.

Infratec (2021) Thermal Imaging > Infrared Cameras > Nyxus Bird LR. https://www.infratec.co.uk/thermography/infrared-camera/nyxus-bird-lr/.

IPBES (2019) Report of the Plenary of the Intergovernmental Science-Policy Platform on Biodiversity and Ecosystem Services on the work of its seventh session. *Plenary of the Intergovernmental Science-Policy Platform on Biodiversity and Ecosystem Services Seventh session*. Paris, 29 April–4 May 2019.

iRed (2021, 19 November) 2021 Emissivity Table for Thermographers. *iRed*. Accessed at https://ired.co.uk/emissivity-table/ (1 May 2022).

Islam, M. M., Ahmed, S. T., Mun, H. S., Bostami, A. B. M. R., Kim, Y. J., and Yang, C. J. (2015) Use of thermal imaging for the early detection of signs of disease in pigs challenged orally with Salmonella typhimurium and Escherichia coli. *African Journal of Microbiology Research* 9(26): 1667–1674. DOI: 10.5897/AJMR2015.7580.

Johnston, D. W., Dale, J., Murray, K. T., Josephson, E., Newton, E., and Wood, S. (2017) Comparing occupied and unoccupied aircraft surveys of wildlife populations: assessing the gray seal (*Halichoerus grypus*) breeding colony on Muskeget Island, USA. *Journal of Unmanned Vehicle Systems* 5(4): 178–191. DOI: 10.1139/juvs-2017-0012.

Joos, E., Giersch, A., Bhatia, K., Heinrich, S. P., Elst, L. T. V., and Kornmeier, J. (2020) Using the perceptual past to predict the perceptual future influences the perceived present – A novel ERP paradigm. *PLoS One* 15(9): e0237663. DOI: 10.1371/journal.pone.0237663.

Jumail, A., Liew, T. S., Salgado-Lynn, M., Fornace, K. M., and Stark, D. J. (2021) A comparative evaluation of thermal camera and visual counting methods for primate census in a riparian forest at the Lower Kinabatangan Wildlife Sanctuary (LKWS), Malaysian Borneo. *Primates* 62(1): 143–151. DOI: 10.1007/s10329-020-00837-y.

Kaldellis, J. K., Apostolou, D., Kapsali, M., and Kondili, E. (2016) Environmental and social footprint of offshore wind energy. Comparison with onshore counterpart. *Renewable Energy* 92: 543–556. DOI: 10.1016/j.renene.2016.02.018.

Kano, F., Hirata, S., Deschner, T., Behringer, V., and Call, J. (2015) Nasal temperature drop in response to a playback of conspecific fights in chimpanzees: a thermos-imaging study. *Physiology and Behavior* 155: 83–94. DOI: 10.1016/j.physbeh.2015.11.029.

Karp, D. (2020) Detecting small and cryptic animals by combining thermography and a wildlife detection dog. *Scientific reports* 10(1): 1–11. DOI: 10.1038/s41598-020-61594-y.

Kays, R., Sheppard, J., Mclean, K., Welch, C., Paunescu, C., Wang, V., Kravit, G., and Crofoot, M. (2019) Hot monkey, cold reality: surveying rainforest canopy mammals using drone-mounted thermal infrared sensors. *International journal of remote sensing* 40(2): 407–419. DOI: 10.1080/01431161.2018.1523580.

Kilgo, J. C., Blake, J. I., Grazia, T. E., Horcher, A., Larsen, M., Mims, T., and Zarnoch, S. J. (2020) Use of roadside deer removal to reduce deer–vehicle collisions. *Human–Wildlife Interactions* 14(1): 13. DOI: 10.26077/t380-nk14.

Kim, M., Chung, O. S., and Lee, J. K. (2021) A Manual for Monitoring Wild Boars (*Sus scrofa*) Using Thermal Infrared Cameras Mounted on an Unmanned Aerial Vehicle (UAV). *Remote Sensing* 13(20): 4141. DOI: 10.3390/rs13204141.

Kingsley, M. C. S., Hammill, M. O., and Kelly, B. P. (1990) Infrared sensing of the under-snow lairs of the ringed seal. *Marine Mammal Science* 6(4): 339–347. DOI: 10.1111/j.1748-7692.1990.tb00363.x.

Kinzie, K., Hale, A., Bennett, V., Romano, B., Skalski, J., Coppinger, K., and Miller, M. F. (2018) *Ultrasonic Bat Deterrent Technology* (No. DOE-GE-07035). General Electric Company, Schenectady, NY (United States). DOI: 10.2172/1484770.

Kirkwood, J. J, and Cartwright, A. (1991) *Behavioral observations in thermal imaging of the big brown bat, Eptesicus fuscus*. SPIE, vol. 1467, Thermosense XIII: 369–371. DOI: 10.1117/12.46448.

Kirkwood, J. J., and Cartwright, A. (1993) Comparison of two systems for viewing bat behavior in the dark. *Proceedings of the Indiana Academy of Science* 102: 133–137.

Kissell, R. E. Jr., and Nimmo, S. K. (2011) A technique to estimate white-tailed deer density using vertical-looking infrared imagery. *Wildlife Biology* 17: 85–92. DOI: 10.2981/10-040.

Kissell, R. E., Jr., and Tappe, P. A. (2004) An assessment of thermal infrared detection rates using white-tailed deer surrogates. *Journal of the Arkansas Academy of Science* 58: 70–73.

Klir, J. J., and Heath, J. E. (1992) An infrared thermographic study of surface temperature in relation to external thermal stress in three species of foxes: The red fox (*Vulpes vulpes*), arctic fox (*Alopex lagopus*), and kit fox (*Vulpes macrotis*). *Physiological Zoology* 65(5): 1011–1021. DOI: 10.1086/physzool.65.5.30158555.

Kuhn, R. A., and Meyer, W. (2009) Infrared thermography of the body surface in the Eurasian otter *Lutra lutra* and the giant otter *Pteronura brasiliensis*. *Aquatic Biology* 6(1–3): 143–152. DOI: 10.3354/ab00176.

Kunz, T. H., and Parsons, S. (2009) *Ecological and Behavioral Methods for the Study of Bats*. 2nd edn. Johns Hopkins University Press.

Lancaster, W. C., Thomson, S. C., Speakman, J. R. (1997) Wing temperature in flying bats measured by infrared thermography. *Journal of Thermal Biology* 22(2): 109–116. DOI: 10.1016/S0306-4565(96)00039-3.

Landry, J. M., Borelli, J. L., and Drouilly, M. (2020) Interactions between livestock guarding dogs and wolves in the southern French Alps. *Journal of Vertebrate Biology* 69(3), 20078.1–18. DOI: /10.25225/jvb.20078.

Lavers, C., Franklin, P., Plowman, A., Sayers, G., Bol, J., Shepard, D., and Fields, D. (2009) Non-destructive high-resolution thermal imaging techniques to evaluate wildlife and delicate biological samples. *Journal of Physics Conference Series* 178: 012040. DOI: 10.1088/1742-6596/178/1/012040.

Lavers, C., Franks, K., Floyd, M., and Plowman, A. (2005) Application of remote thermal imaging and night vision technology to improve endangered wildlife resource management with minimal animal distress and hazard to humans. *Journal of Physics: Conference Series* 15(16): 207–212. DOI: 10.1088/1742-6596/15/1/035.

Lazarevic, L. (2009) Improving the efficiency and accuracy of nocturnal bird surveys through equipment selection and partial automation. Unpublished PhD thesis Brunel University, London, UK.

Lee, S., Song, Y., and Kil, S. H. (2021) Feasibility analyses of real-time detection of wildlife using UAV-derived thermal and rgb images. *Remote Sensing* 13(11): 2169. DOI: 10.3390/rs13112169.

Lethbridge, M. R., Stead, M. G., Wells, C., and Shute, E. (2020) *Report of state-wide census of wild fallow deer in Tasmania project: Part A: Baseline aerial survey of fallow deer population, central and north-eastern Tasmania*. Department of Primary Industries, Parks, Water and Environment.

Lethbridge, M., Stead, M., and Wells, C. (2020) Estimating kangaroo density by aerial survey: a comparison of thermal cameras with human observers. *Wildlife Research* 46(8): 639–648. DOI: 10.1071/WR18122.

Lhoest, S., Linchant, J., Quevauvillers, S., Vermeulen, C., and Lejeune, P. (2015) How Many Hippos (Homhip): Algorithm for Automatic Counts of Animals with Infra-Red Thermal Imagery from UAV. *International Archives of the Photogrammetry, Remote Sensing and Spatial Information Sciences* XL-3/W3: 355–362. DOI: 10.5194/isprsarchives-XL-3-W3-355-2015.

Liechti, F., Peter, D., and Komenda-Zehnder, S. (2003) Nocturnal bird migration in Mauritania – first records. *Journal für Ornithologie* 144: 445–450. DOI: 10.1007/BF02465507.

Liechti, F., Bruderer, B., and Paproth, H. (1995) Quantification of nocturnal bird migration by moonwatching: comparison with radar and infrared observations. *Journal of Field Ornithology* 66(4): 457–468.

Locke, S. L., Lopez, R. R., Peterson, M. J., Silvy, N. J., and Schwertner, T. W. (2006) Evaluation of portable infrared cameras for detecting Rio Grande wild turkeys. *Wildlife Society Bulletin* 34(3): 839–844. DOI: 10.2193/0091-7648(2006)34[839:EOPICF]2.0.CO;2.

Longmore, S. N., Collins, R. P., Pfeifer, S., Fox, S. E., Mulero-Pazmany, M., Bezombes, F., Goodwin, A., De Juan Ovelar, M., Knapen, J. H., and Wich, S. A. (2017) Adapting astronomical source detection software to help detect animals in thermal images obtained by unmanned aerial systems. *International Journal of Remote Sensing* 38: 2623–2638. DOI: 10.48550/arXiv.1701.01611.

Mammeri A., Zhou, D., and Boukerche, A. (2016) Animal–vehicle collision mitigation system for automated vehicles. *IEEE Transactions on Systems, Man, and Cybernetics: Systems* 46(9): 1287–1299. DOI: 10.1109/TSMC.2015.2497235.

Marini, F., Franzetti, B., Calabrese, A., Cappellini, S., and Focardi, S. (2009) Response to human presence during nocturnal line transect surveys in fallow deer (*Dama dama*) and wild boar (*Sus scrofa*). *European Journal of Wildlife Research* 55(2): 107–115. DOI: 10.1007/s10344-008-0222-7.

Martin, J. M., and Barboza, P. S. (2020) Thermal biology and growth of bison (*Bison bison*) along the Great Plains: examining four theories of endotherm body size. *Ecosphere* 11(7): e03176. DOI: 10.1002/ecs2.3176.

Matzner, S., Cullinan, V. I., and Duberstein, C. A. (2015) Two-dimensional thermal video analysis of offshore bird and bat flight. *Ecological Informatics* 30: 20–28. DOI: 10.1016/j.ecoinf.2015.09.001.

Matzner, S., Warfel, T. and Hull, R. (2020) ThermalTracker-3D: A thermal stereo vision system for quantifying bird and bat activity at offshore wind sites. *Ecological Informatics* 57: 101069. DOI: 10.1016/j.ecoinf.2020.101069.

Maxwell, S. L., Fuller, R. A., Brooks, T. M. and Watson, J. E. M. (2016) Biodiversity: The ravages of guns, nets and bulldozers. *Nature* 536: 143–145. DOI: 10.1038/536143a.

Mazor, T., Doropoulos, C., Schwarzmueller, F., Gladish, D. W., Kumaran, G., Merkel, K., Marco, M. D., and Gagic, V. (2018) Global mismatch of policy and research on drivers of biodiversity loss. *Nature Ecology & Evolution* 2: 1071–1074. DOI: 10.1038/s41559-018-0563-x.

McCafferty, D. J., Moncrieff, J. B., Taylor, I. R., and Boddie, G. F. (1998) The use of IR thermography to measure the radiative temperature and heat loss of a barn owl (*Tyto alba*). *Journal of Thermal Biology* 23(5): 311–318. DOI: 10.1016/S0306-4565(98)00022-9.

McCafferty, D. J. (2007) The value of infrared thermography for research on mammals: previous applications and future directions. *Mammal Review* 37(3), 207–223. DOI: 10.1111/j.1365-2907.2007.00111.x.

McCarthy, E. D., Martin, J. M., Boer, M. M., and Welbergen, J. A. (2021) Drone-based thermal remote sensing provides an effective new tool for monitoring the abundance of roosting fruit bats. *Remote Sensing in Ecology and Conservation* 7(3): 461–474. DOI: 10.1002/rse2.202.

McCollister, M. F. and Van Manen, F. T. (2010) Effectiveness of Wildlife Underpasses and Fencing to Reduce Wildlife-Vehicle Collisions. *The Journal of Wildlife Management* 74 (8): 1722–1731. DOI: 10.2193/2009-535.

McGowan, N. E., Scantlebury, D. M., Maule, A. G., and Marks, N. J. (2018) Measuring the emissivity of mammal pelage. *Quantitative InfraRed Thermography Journal* 15(2): 214–222. DOI: 10.1080/17686733.2018.1437239.

McGregor, H., Moseby, K., Johnson, C. N., and Legge, S. (2021) Effectiveness of thermal cameras compared to spotlights for counts of arid zone mammals across a range of ambient temperatures. *Australian Mammalogy* 44(1): 59–66. DOI: 10.1071/AM20040.

McMahon, M. C., Ditmer, M. A., Isaac, E. J., Moore, S. A., and Forester, J. D. (2021) Evaluating Unmanned Aerial Systems for the Detection and Monitoring of Moose in Northeastern Minnesota. *Wildlife Society Bulletin* 45(2): 312–324. DOI: 10.1002/wsb.1167.

Medolago, C. A. B., Abra, F. D., and Prist, P. R. (2021) Use of a portable thermograph as a potential tool to identify nocturnal airport bird risks. *Brazilian Journal of Animal and Environmental Research* 4(2): 2360–2370. DOI: 10.34188/bjaerv4n2-065.

Metz, I. C., Ellerbroek, J., Mühlausen, T., Kügler, D., and Hoekstra, J. M. (2020) The Bird Strike Challenge. *Aerospace* 7 (3): 26. DOI: 10.3390/aerospace7030026.

Miard, P. (2020) Distribution, methodological validation and ecology of nocturnal island mammals in peninsular Malaysia. Doctoral Thesis. Universiti Sains Malaysia.

Millette, T. L., Slaymaker, D., Marcano, E., Alexander, C., and Richardson, L. (2011) AIMS-Thermal – a Thermal and High Resolution Color Camera System Integrated with GIS for Aerial Moose and Deer Census in Northeastern Vermont. *Alces* 47: 27–37.

Mills, W. E., Harrigal, D. E., Owen, S. F., Dukes, W. F., Barrineau, D. A. and Wiggers, E. P. (2011) Capturing clapper rails using thermal imaging technology. *Journal of Wildlife Management* 75(5): 1218–1221. DOI: 10.1002/jwmg.142.

Mitchell, W. F. and Clarke, R. H. (2019) Using infrared thermography to detect night-roosting birds. *Journal of Field Ornithology* 90: 39–51. DOI: 10.1111/jofo.12285.

Montgomery, R. A. (2020) Poaching is Not One Big Thing. *Trends in Ecology & Evolution* 35(6): 472–475. DOI: 10.1016/j.tree.2020.02.013.

Moore, G. A. (1991) *Crossing the Chasm: Marketing and Selling High-tech Products to Mainstream Customers.* Harper Business Essentials.

Morelle, K., Bouche, P., Lehaire, F., Leeman, V., and Lejeune, P. (2012) Game species monitoring using road-based distance sampling in association with thermal imagers: a covariate analysis. *Animal Biodiversity and Conservation* 35(2): 253–265. DOI: 10.32800/abc.2012.35.0253.

Mulero-Pazmany, M., Stopler, R., van Essen, L. D., Negro, J. J., and Sassen, T. (2014) Remotely piloted aircraft systems as a rhinoceros anti-poaching tool in Africa. *PLoS ONE* 9(1): e83873. DOI: 10.1371/journal.pone.0083873.

Munshi-South, J., and Wilkinson, G. S. (2010) Bats and birds: exceptional longevity despite high metabolic rates. *Ageing Research Reviews* 9(1): 12–19.

Mutalib, A. H. A., Ruppert, N., Kamaruszaman, S. A., Jamsari, F. F., and Rosely, N. F. N. (2019) Feasibility of thermal imaging using unmanned aerial vehicles to detect Bornean orangutans. *Journal of Sustainability Science and Management* 14(5): 182–194.

Mysłajek, R. W., Olkowska, E., Wronka-Tomulewicz, M., and Nowak, S. (2020) Mammal use of wildlife crossing structures along a new motorway in an area recently recolonized by wolves. *European Journal of Wildlife Research* 66: 79. DOI: 10.1007/s10344-020-01412-y.

Nääs, I. A., Garcia, R. G., and Caldara, F. R. (2014) Infrared thermal image for assessing animal health and welfare. *Journal of Animal Behaviour and Biometeorology* 2(3): 66–72. DOI: 10.14269/2318-1265/jabb.v2n3p66-72.

Narayan, E., Perakis, A., and Meikle, W. (2019) Using Thermal Imaging to Monitor Body Temperature of Koalas (*Phascolarctos cinereus*) in A Zoo Setting. *Animals* 9(12): 1094. DOI: 10.3390/ani9121094.

Naugle, D. E., Jenks, J. A., and Kernohan, B. J. (1996) Use of thermal infrared sensing to estimate density of white-tailed deer. *Wildlife Society Bulletin* 24: 37–43.

Netflix (2020) *Night on Earth.* Accessed at https://www.netflix.com/watch/80988075?trackId=255824129.

Newson, S. E., Evans, H. E., Gillings, S., Jarrett, D., Raynor, R., and Wilson, M. W. (2017). Large-scale citizen science improves assessment of risk posed by wind farms to bats in southern Scotland. *Biological Conservation*, 215, 61–71. DOI: 10.1016/j.biocon.2017.09.004.

Nottingham, C. M., Glen, A. S., and Stanley, M. C. (2019) Snacks in the city: the diet of hedgehogs in Auckland urban forest fragments. *New Zealand Journal of Ecology* 43(2): 3374. DOI: 10.20417/nzjecol.43.24.

Nyhus, P. J. (2016) Human–Wildlife Conflict and Coexistence. *Annual Review of Environment and Resources* 41: 143–171. DOI: 10.1146/annurev-environ-110615-085634.

O'Neal, B. J., Stafford, J. D., and Larkin, R. P. (2010) Waterfowl on weather radar: applying ground-truth to classify and quantify bird movements. *Journal of Field Ornithology* 81: 71–82. DOI: 10.1111/j.1557-9263.2009.00263.x.

Öhman, C. (2014) Measurement in thermography. Flir Systems and The Infrared Training Center.

Oishi, Y., Oguma, H., Tamura, A., Nakamura, R., and Matsunaga, T. (2018) Animal Detection Using Thermal Images and Its Required Observation Conditions. *Remote Sensing* 10(7): 1050. DOI: 10.3390/rs10071050.

Ono, M., Igarashi, T., Ohno, E., and Sasaki, M. (1995) Unusual thermal defense by a honeybee against mass attack by hornets. *Nature Letters* 337: 334–336. DOI: 10.1038/377334a0.

Otálora-Ardila, A., Torres, J. M., Barbier, E., Pimentel, N. T., Barbosa Leal, E. S., and Bernard, E. (2019) Thermally-assisted monitoring of bat abundance in an exceptional cave in Brazil's Caatinga drylands. *Acta Chiropterologica* 21(2): 411–423. DOI: 10.3161/15081109ACC2019.21.2.016.

Ovadia, O., Pinshow, B., and Lotem, A. (2002) Thermal imaging of House Sparrow nestlings: the effect of begging behavior and nestling rank. *Condor* 104(4): 837–842. DOI: 10.1093/condor/104.4.837.

Parker, H. D. Jr., and Driscoll, R. S. (1972) An Experiment in Deer Detection by Thermal Scanning. *Journal of Range Management* 25(6): 480–481. DOI: 10.2307/3897015.

Perryman, W. L., Donahue, M. A., Laake, J. L., and Martin, T. E. (1999) Diel variation in migration rates of eastern pacific gray whales measured with thermal imaging sensors. *Marine Mammal Science* 15(2): 426–445. DOI: 10.1111/j.1748-7692.1999.tb00811.x.

Phillips, P. K. and Heath, J. E. (1992) Heat exchange by the pinna of the African Elephant (Loxodonta africana). *Comparative Biochemistry and Physiology, Comparative Physiology* 101A(4): 693–699. DOI: 10.1016/0300-9629(92)90345-q.

Playà-Montmany, N., and Tattersall, G. J. (2021) Spot size, distance and emissivity errors in field applications of infrared thermography. *Methods in Ecology and Evolution* 12(5): 828–840. DOI: 10.1111/2041-210X.13563.

Polat, B., Colak, A., Cengiz, M., Yanmaz, L. E., Oral, H., Bastan, A., Kaya, S., and Hayirli, A. (2010) Sensitivity and specificity of infrared thermography in detection of subclinical mastitis in dairy cows. *Journal of Dairy Science* 93(8): 3 525–532. DOI: 10.3168/jds.2009-2807.

Potvin, F., and Breton, L. (2005) Testing 2 aerial survey techniques on deer in fenced enclosures: visual double-counts and thermal infrared sensing. *Wildlife Society Bulletin* 33: 317–325. DOI: 10.2193/0091-7648(2005)33[317:FTFTAS]2.0.CO;2.

Preston, T. M., Wildhaber, M. L., Green, N. S., Albers, J. L., and Debenedetto, G. P. (2021) Enumerating White-Tailed Deer Using Unmanned Aerial Vehicles. *Wildlife Society Bulletin* 45(1): 97–108. DOI: 10.1002/wsb.1149.

Project Splatter (2020) *Wildlife roadkill map: 2019*. Accessed at https://projectsplatter.co.uk/2020/04/07/wildlife-roadkill-map-2019/ (26 May 2022).

Prosekov, A. Y. (2020) Introduction of digital technologies in methods of accounting for hunting animals. *Proceedings of the Lower Volga Agro-University Comp* 3(59): 268–274. DOI: 10.32786/2071-9485-2020-03-28.

Psiroukis, V., Malounas, I., Mylonas, N., Grivakis, K. E., Fountas, S., and Hadjigeorgiou, I. (2021) Monitoring of free-range rabbits using aerial thermal imaging. *Smart Agricultural Technology* 1: 100002. DOI: 10.1016/j.atech.2021.100002.

Rahman, D. A., and Rahman, A. A. A. F. (2021) Performance of unmanned aerial vehicle with thermal imaging, camera trap, and transect survey for monitoring of wildlife. *IOP Conference Series: Earth and Environmental Science* 771(1): 012011. DOI: 10.1088/1755-1315/771/1/012011.

Rahman, D. A., Setiawan, Y., Wijayanto, A. K., Rahman, A. A. A. F., and Martiyani, T. R. (2020) An experimental approach to exploring the feasibility of unmanned aerial vehicle and thermal imaging in terrestrial and arboreal mammals research. *In E3S Web of Conferences* 211: 02010. DOI: 10.1051/e3sconf/202021102010.

Ramin, C., Devore, E. E., Wang, W., Pierre-Paul, J., Wegrzyn, L. R., and Schernhammer, E. S. (2015) Night shift work at specific age ranges and chronic disease risk factors. *Occupational and Environmental Medicine* 72 (2): 100–107. DOI: 10.1136/oemed-2014-102292.

Reichard, J. D., Prajapati, S. I., Austad, S. N., Keller, C., and Kunz, T. H. (2010) Thermal windows on Brazilian free-tailed bats facilitate thermoregulation during prolonged flight. *Integrative and Comparative Biology* 50(3): 358–370. DOI: 10.1093/icb/icq033.

Ritchie, H. and Roser, M. (2021) Biodiversity. *Our World in Data*. Accessed at https://ourworldindata.org/biodiversity (17 May 2022).

Roberts, B. R., and Osborne, J. L. (2019) Testing the efficacy of a thermal camera as a search tool for locating wild bumble bee nests. *Journal of Apicultural Research* 58(4): 494–500. DOI: 10.1080/00218839.2019.1614724.

Robinson, R., Smith, T. S., Larson, R. T., and Kirschhoffer, B. J. (2014) Factors influencing the efficacy of forward-looking infrared in polar bear den detection. *BioScience* 64(8): 735–742. DOI: 10.1093/biosci/biu095.

Sabol, B. M., and Hudson, M. K. (1995) Technique Using Thermal Infrared-Imaging for Estimating Populations of Gray Bats. *Journal of Mammalogy* 76(4): 1242–1248. DOI: 10.2307/1382618.

Santangeli, A., Chen, Y., Kluen, E., Chirumamilla, R., Tiainen, J., and Loehr, J. (2020) Integrating drone-borne thermal imaging with artificial intelligence to locate bird nests on agricultural land. *Scientific reports* 10(1), 1–8. DOI: 10.1038/s41598-020-67898-3.

Sasse, D. B. (2003). Job-Related Mortality of Wildlife Workers in the United States, 1937–2000. *Wildlife Society Bulletin* (1973–2006), 31(4), 1015–1020. http://www.jstor.org/stable/3784446.

Schaefer, A. L., Cook, N. J., Bench, C., Chabot, J. B., Colyn, J., Liu, T., Okine, T. E. K., Stewart, M., Webster, J. R. (2012) The non-invasive and automated detection of bovine respiratory disease onset in receiver calves using infrared thermography. *Research in Veterinary Science* 93(2): 928–935. DOI: 10.1016/j.rvsc.2011.09.021.

Schedl, D. C., Kurmi, I., and Bimber, O. (2020) Airborne optical sectioning for nesting observation. *Scientific reports* 10(1): 7254. DOI: 10.1038/s41598-020-63317-9.

Schirmacher, M. R. (2020) Evaluating the effectiveness of an ultrasonic acoustic deterrent in reducing bat fatalities at wind energy facilities (No. DOE-BCI-0007036). *U.S. Department of Energy: Office of Scientific and Technical Information*. DOI: 10.2172/1605929.

Scholten, C. N., A. J. Kamphuis, K. J. Vredevoodg, K. G. Lee-Strydhorst, J. L. Atma, C. B. Shea, O. N. Lamberg, and D. S. Proppe (2019) Real-time thermal imagery from an unmanned aerial vehicle can locate ground nests of a grassland songbird at rates similar to traditional methods. *Biological Conservation* 233: 241–246. DOI: 10.1016/j.biocon.2019.03.001.

Seier, G., Hödl, C., Abermann, J., Schöttl, S., Maringer, A., Hofstadler, D. N., Pröbstl-Haider, U., and Lieb, G. K. (2021) Unmanned aircraft systems for protected areas: gadgetry or necessity? *Journal for Nature Conservation*, 126078. DOI: 10.1016/j.jnc.2021.126078.

Semel, B. P., Karpanty, S. M., Vololonirina, F. F., and Rakotonanahary, A. N. (2019) Eyes in the sky: Assessing the feasibility of low-cost, ready-to-use Unmanned Aerial Vehicles to monitor primate populations directly. *Folia Primatologica* 91(1): 69–82. DOI: 10.1159/000496971.

Seymour, A. C., Dale, J., Hammill, M., Halpin, P. N., and Johnston, D. W. (2017) Automated detection and enumeration of marine wildlife using unmanned aircraft systems (UAS) and thermal imagery. *Scientific reports* 7(1): 45127. DOI: 10.1038/srep45127.

Shaffer, L. J., Khadka, K. K., Van Den Hoek, J., and Naithani, K. J. (2019) Human–elephant conflict: A review of current management strategies and future directions. *Frontiers in Ecology and Evolution*, 6, 235. DOI: 10.3389/fevo.2018.00235.

Shewring, M. P. and Vafidis, J. O. (2020) Using UAV-mounted thermal cameras to detect the presence of nesting nightjar in upland clear-fell: A case study in South Wales, UK. *Ecological Solutions and Evidence* 2(1): e12052. DOI: 10.1002/2688-8319.12052.

Shirai, M., Sugimoto, T., Ishino, R., and Kado, H. (2020) The Effectiveness of Visual Scaring Techniques Against Grey Herons, Ardea cinerea. In *Proceedings of the Vertebrate Pest Conference* 29.

Sidle, J. G., Nagel, H. G., Clarke, R., Gilbert, C., Stuart, D., Willburn, K., and Orr, M. (1993) Aerial thermal infrared imaging of sandhill cranes on the Platte river, Nebraska. *Remote Sensing of Environment* 43(3): 333–341. DOI: 10.1016/0034-4257(93)90074-8.

Simmons, J. A., Brown, P. E., Vargas-Irwin, C. E., and Simmons, A. M. (2020) Big brown bats are challenged by acoustically-guided flights through a circular tunnel of hoops. *Scientific reports* 10(1); 832. DOI: 10.1038/s41598-020-57632-4.

Sliwinski, K., Strauß, E., Jung, K., and Siebert, U. (2021) Comparison of spotlighting monitoring data of European brown hare (*Lepus europaeus*) relative population densities with infrared thermography in agricultural landscapes in Northern Germany. *PloS one* 16(7): e0254084. DOI: 10.1371/journal.pone.0254084.

Smallwood, K. S. and Bell, D. A. (2020) Relating bat passage rates to wind turbine fatalities. *Diversity* 12(2): 84. DOI: 10.3390/d12020084.

Smallwood, K. S., Bell, D. A., and Standish, S. (2020) Dogs detect larger wind energy effects on bats and birds. *The Journal of Wildlife Management* 84(5): 852–864. DOI: 10.1002/jwmg.21863.

Smallwood, K. S. and Bell, D. A. (2020) Effects of Wind Turbine Curtailment on Bird and Bat Fatalities. *The Journal of Wildlife Management* 84(4): 685–696. DOI: 10.1002/jwmg.21844.

Smart, J. C., Ward, A. I., and White, P. C. L. (2004) Monitoring woodland deer populations in the UK: an imprecise science. *Mammal Review* 34(1–2): 99–114. DOI: 10.1046/j.0305-1838.2003.00026.x.

Smith, T. S., Amstrup, S. C., Kirschhoffer, B. J., and York, G. (2019) Efficacy of aerial forward-looking infrared surveys for detecting polar bear maternal dens. *PLoS ONE* 15(2): e0222744. DOI: 10.1371/journal.pone.0222744.

Smith, D. J., van der Ree, R., and Rosell, C. (2015) Wildlife Crossing Structures: An Effective Strategy to Restore or Maintain Wildlife Connectivity Across Roads in van Der See, R., Smith, D. J., and Grilo, C. (Eds) *Handbook of Road Ecology*. John Wiley & Sons Ltd, Chichester. Pp 172–183.

Soroko, M. and Morel, M. C. G. (2016) *Equine Thermography in Practice*. CABI Publishing, Oxford.

South, K. E., Haynes, K., and Jackson, A. C. (2020) Diagnosis of hypothermia in the European hedgehog, Erinaceus europaeus, using infrared thermography. *Journal of Thermal Biology* 90: 102574. DOI: 10.1016/j.jtherbio.2020.102574.

Spaan, D., Burke, C., McAree, O., Aureli, F., Rangel-Rivera, C. E., Hutschen-Reiter, A., Longmore, S. N., McWhirter, P. R. and Wich, S. A. (2019) Thermal infrared imaging from drones offers a major advance for spider monkey surveys. *Drones* 3(2): 34. DOI: 10.3390/drones3020034.

Steen, K. A., Villa-Henriksen, A., Therkildsen, O. R., and Green, O. (2012) Automatic Detection of Animals in Mowing Operations Using Thermal Cameras. *Sensors* 12(6): 7587–7597. DOI: 10.3390/s120607587.

Stephenson, M. D., Schulte, L. A., and Klaver, R. W. (2019) Quantifying thermal-imager effectiveness for detecting bird nests on farms. *Wildlife Society Bulletin* 43(2): 302–307. DOI: 10.1002/wsb.962.

Stewart, G. and Coles, C. (2007) Poor evidence-base for assessment of windfarm impacts on birds. *Environmental Conservation* 34(1): 1–11. DOI: 10.1017/S0376892907003554.

Storm, D. J., Samuel, M. D., Van Deelen, T. R., Malcolm, K. D., Rolley, R. E., Frost, N. A., Bates, D. P., and Richards, B. J. (2011) Comparison of visual-based helicopter and fixed wing forward looking infrared surveys for counting white-tailed deer *Odocoileus virginianus*. *Wildlife Biology* 17(4): 431–440. DOI: 10.2981/10-062.

Sumbera, R., Zelova, J., Kunc, P., Knizkova, I., and Burda, H. (2007) Patterns of surface temperatures in two male-rats (*Bathyergidae*) with different social system as revealed by IR-thermography. *Physiology & Behaviour* 92(3): 526–532. DOI: 10.1016/j.physbeh.2007.04.029.

Syposz, M., Padget, O., Willis, J., Van Doren, B. M., Gillies, N., Fayet, A. L., Wood, M. J., Alejo, A., and Guilford, T. (2021) Avoidance of different durations, colours and intensities of artificial light by adult seabirds. *Scientific reports* 11(1): 18941. DOI: 10.1038/s41598-021-97986-x.

Tattersall, G. J. (2016) Infrared thermography: A non-invasive window into thermal physiology. In *Comparative Biochemistry and Physiology, Part A*. DOI: 10.1016/j.cbpa.2016.02.02.

Tattersall, G. J., Andrade, D. V., and Abe, A. S. (2009) Heat Exchange from the Toucan Bill Reveals a Controllable Vascular Thermal Radiator. *Science* 325(5939): 468–470. DOI: 10.1126/science.1175553.

Tattersall, G. J., and Cadena, V. (2010) Insights into animal temperature adaptations revealed through thermal imaging. *The Imaging Science Journal* 58: 261–268. DOI: 10.1179/136821910X12695060594165.

The Deer Initiative (2008) Records & Surveys: Night Census. *The Deer Initiative*. Accessed at https://www.thedeerinitiative.co.uk/uploads/guides/105.pdf (24 May 2022).

Theriault, D. H., Fuller, N. W., Jackson, B. E., Bluhm, E., Evangelista, D., Wu, K., Betke, M., and Hedrick, T. L. (2014) A protocol and calibration method for accurate multi-camera field videography. *Journal of Experimental Biology* 217(11): 1843–1848. DOI: 10.1242/jeb.100529.

Torgersen, C. E., Price, D. M., Li, H. W., and McIntosh, B. A. (1995) Thermal refugia and chinook salmon habitat in Oregon: Applications of airborne thermal videography. In: Proceedings of the 15th Biennial Workshop on Color Photography and Videography (P. Mausel, ed.) *American Society for Photogrammetry and Remote Sensing*, Terre Haute, IN. Pp 167–171.

Tuneu-Corral, C., Puig-Montserrat, X., Flaquer, C., Mas, M., Budinski, I., and López-Baucells, A. (2020) Ecological indices in long-term acoustic bat surveys for assessing and monitoring bats' responses to climatic and land-cover changes. *Ecological Indicators* 110: 105849. DOI: 10.1016/j.ecolind.2019.105849.

Udevitz, M. S., Burn, D. M., and Webber, M. A. (2008) Estimation of walrus populations on sea ice with infrared imagery and aerial photography. *Marine Mammal Science* 24: 57–70. DOI: 10.1111/J.1748-7692.2007.00169.X.

University of Glasgow (n.d.) Behavioural Thermoregulation. *University of Glasgow*. Accessed at https://www.gla.ac.uk/researchinstitutes/bahcm/research/sigs/thermalecologygroup/behaviouralthermoregulation/ (31 May 2022).

Vinson, S. G., Johnson, A. P., and Mikac, K. M. (2020) Thermal cameras as a survey method for Australian arboreal mammals: a focus on the greater glider. *Australian Mammalogy* 42(3): 367–374. DOI: 10.1071/AM19051.

Voigt, U. and Seibert, U. (2020) Survival rates on pre-weaning European hares (*Lepus europaeus*) in an intensively used agricultural area. *European Journal of Wildlife Research* 66: 67. DOI: 10.1007/s10344-020-01403-z.

von Schweinitz, D. G. (1999) Thermographic diagnostics in equine back pain. *Veterinary Clinics of North America: Equine Practice* 15(1): 161–177. DOI: 10.1016/s0749-0739(17)30170-0.

Ward, S., Hensler, J., Alsalam, B., and Gonzalez, L. F. (2016) Autonomous UAVs wildlife detection using thermal imaging, predictive navigation and computer vision. In *2016 IEEE Aerospace Conference* (pp. 1–8). IEEE. DOI: 10.1109/AERO.2016.7500671.

Weissenböck, N. M., Weiss, C. M., Schwammer, H. M., and Kratochvil, H. (2010) Thermal windows on the body surface of African elephants (*Loxodonta africana*) studied by infrared thermography. *Journal of Thermal Biology* 35: 182–188. DOI: 10.1016/j.jtherbio.2010.03.002.

West Midlands Bird Ringing Group (2020) *Getting the most from your thermal camera a guide to using a thermal image camera for ringing and surveying PDF*. Accessed at https://www.westmidlandsringinggroup.co.uk/thermal-birding (10 May 2021).

Wiggers, E. P., and Beckerman, S. F. (1993) Use of thermal infrared sensing to survey white-tailed deer populations. *Wildlife Society Bulletin* 21(3): 263–268.

Williamson, A., Lombardi, D. A., Folkard, S., Stutts, J., Courtney, T. K., and Connor, J. L. (2011) The link between fatigue and safety. *Accident Analysis and Protection* 43(2): 498–515. DOI: 10.1016/j.aap.2009.11.011.

Willis, K., Horning, M., Rosen, D. A. S., and Trites, A. W. (2005) Spatial variation of heat flux in Stellar sea lions: evidence for consistent avenues of heat exchange along the body trunk. *Journal of Experimental Marine Biology and Ecology* 315(2): 163–175. DOI: 10.1016/j.jembe.2004.09.018.

Witczuk, J., Pagacz, S., Zmarz, A., and Cypel, M. (2018) Exploring the feasibility of unmanned aerial vehicles and thermal imaging for ungulate surveys in forests-preliminary results. *International Journal of Remote Sensing* 39(15–16): 5504–5521. DOI: 10.1080/01431161.2017.1390621.

Witt, R. R., Beranek, C. T., Howell, L. G., Ryan, S. A., Clulow, J., Jordan, N. R., Denholm, B., and Roff, A. (2020) Real-time drone derived thermal imagery outperforms traditional survey methods for an arboreal forest mammal. *Plos one* 15(11): e0242204. DOI: 10.1371/journal.pone.0242204.

Wride, M. C., and Baker, K. (1977) Thermal Imagery for Census of Ungulates (NASA). In *Proceedings of the 11th International Symposium on Remote Sensing of Environment*, 2.

Wright, A. J., Araújo-Wang, C., Wang, J. Y., Ross, P. S., Tougaard, J., Winkler, R., Márquez, M. C., Robertson, F. C., Fawcett Williams, K., and Reeves, R. R. (2020) How 'Blue' is 'Green' Energy? *Trends in Ecology and Evolution* 35(3): 235–244. DOI: 10.1016/j.tree.2019.11.002.

Wu, Z., Fuller, N., Theriault, D., and Betke, M. (2014) A thermal infrared video benchmark for visual analysis. In *Proceedings of the IEEE Conference on Computer Vision and Pattern Recognition Workshops*: 201–208. DOI: 10.1109/CVPRW.2014.39.

WWF (n.d.) Oil and gas development. *World Wildlife Fund Inc.* Accessed at https://www.worldwildlife.org/threats/oil-and-gas-development#:~:text=Specifically%2C%20oil%20and%20gas%20exploration,who%20depend%20on%20these%20ecosystems (24 May 2022).

Yang, X., Schaaf, C., Strahler, A., Kunz, T., Fuller, N., Betke, M., Wu, Z., Wang, Z., Theriault, D., Culvenor, D., Jupp, D., Newnham, G., and Lovell, J. (2013) Study of bat flight behavior by combining thermal image analysis with a LiDAR forest reconstruction. *Canadian Journal of Remote Sensing* 39(sup1): S112–S125. DOI: 10.5589/m13-034.

York, G., Amstrup, S. C., and Simac, K. (2004) *Using forward looking infrared (FLIR) imagery to detect polar bear maternal dens: operations manual*. US Department of the Interior, Minerals Management Service.

Zainalabidin, D. N. U. A. P., Miard, P., and Grafe, T. U. (2020) Distribution of arboreal nocturnal mammals in northern Borneo. *Scientia Bruneiana* 19(1): 78–85. DOI: 10.46537/scibru.v19i1.120.

Zehnder, S., Akesson, S., Liechti, F., and Bruderer, B. (2001) Nocturnal autumn bird migration at Falsterbo, south Sweden. *Journal of Avian Biology* 32(3): 239–248. DOI: 10.1111/j.0908-8857.2001.320306.x.

Zhou, D., Wang, J., and Wang, S. (2012) Contour based HOG deer detection in thermal images for traffic safety. In *Proceedings of the International Conference on Image Processing, Computer Vision, and Pattern Recognition* (IPCV) (p. 1). The Steering Committee of The World Congress in Computer Science, Computer Engineering and Applied Computing (WorldComp).

Zini, V., Wäber, K., Hornigold, K., Lake, I., and Dolman, P. M. (2021) Human and environmental associates of local species-specific abundance in a multi-species deer assemblage. *European Journal of Wildlife Research* 67(6): 99. DOI: 10.1007/s10344-021-01539-6.

Zitterbart, D. P., Kindermann, L., Burkhardt, E., and Boebel, O. (2013) Automatic Round-the-Clock Detection of Whales for Mitigation from Underwater Noise Impacts. *PLoS ONE* 8(8): e71217. DOI: 10.1371/journal.pone.0071217.

Zukal, J., Pikura, J., and Bandouchova, H. (2015) Bats as bioindicators of heavy metal pollution: history and prospect. *Mammalian Biology* 80: 220–227. DOI: 10.1016/j.mambio.2015.01.001.

Zurich (2021) *Staycation boom drives 54% increase in wildlife fatalities on UK roads*. Accessed at https://www.zurich.co.uk/news-and-insight/staycation-boom-drives-54-per-cent-increase-in-wildlife-fatalities-on-uk-roads (26 May 2022).

Index

Printed in the USA
CPSIA information can be obtained
at www.ICGtesting.com
JSHW052051011223
52894JS00001B/2

9 781784 273873